T0256371

THE COGNITIVE ARTIFACTS
OF DESIGNING

THE COGNITIVE ARTIFACTS OF DESIGNING

Willemien Visser
*INRIA – National Institute for Research
in Computer Science and Control*

2006

LAWRENCE ERLBAUM ASSOCIATES, PUBLISHERS
Mahwah, New Jersey   London

Camera ready copy for this book was provided by the author.

Lawrence Erlbaum Associates, Inc., Publishers
10 Industrial Avenue
Mahwah, New Jersey 07430
www.erlbaum.com

Cover design by Kathryn Houghtaling Lacey

Library of Congress Cataloging-in-Publication Data

Visser, Willemien.
    The cognitive artifacts of designing / Willemien Visser.
        p. cm.
Includes bibliographical references and index.
ISBN 0-8058-5511-4 (cloth : alk. paper)
1. Human engineering. 2. Industrial design. 3. Design—Human factors. 4. Cognitive science. I. Title.
TA166.V575 2006
745.2—dc22   2006009363
                                                                    CIP

Books published by Lawrence Erlbaum Associates are printed on acid-free paper, and their bindings are chosen for strength and durability.

Printed in the United States of America
10 9 8 7 6 5 4 3 2 1

For my mother,
Helena,
Mara, Siebrich, Geerke and Willemijn
—and, of course, my father,
who, unwittingly but nonetheless with great wit and wisdom,
taught me to design my own path through life

Contents in Brief

Contents

Foreword

Design is one of the most characteristic of human endeavors. The history of our species is one of reshaping the environment to reflect our visions and to better meet our needs. Today, more than ever before, we make our world through our designs. This is not to assert that humans are always, or even typically, wise designers who create good designs. Sadly, I think that is not true. It is just to say that we are characteristically designers.

Design would accordingly seem to be a core topic, perhaps *the* core topic, in cognitive science, that is, in the interdisciplinary study of the complex intellectual capabilities that inhere in being human. One of the key insights that shaped the past quarter-century of research in cognitive science was the recognition that complex human activity is highly situated: Human thought, action, and experience are modulated by the social and material contexts in which they occur. Indeed, human activity is minutely shaped by context moment by moment as it occurs.

Design is a compelling example of how complex intellectual activity requires multidisciplinary description referenced to context-specific knowledge and cognitive skill. And indeed, understanding design is a preeminent contemporary concern in a variety of complex domains including software engineering, programming, and algorithm design, mechanical systems and engineering design, architecture and environmental design, maps, routes and other geospatial design, designs of policies and social protocols, and the design of meals, fabrics, and other domestic artifacts.

Given the relationship of design to being human and to many, if not all, critical domains of human activity, it is not surprising that understanding design has attracted some of the finest minds of our era. For example, today, I think, we are still assimilating—and reconciling—the powerful visions of Herbert Simon and Donald Schön articulated during the last four decades. In just this regard, it *is* surprising however that design is not a better-institutionalized area of study within what might be called the academic establishment of cognitive science. There is a curiously yawning gap between the theorizing of giants like Simon and Schön, and the research agendas, dissertations, and journal papers of rank-and-file cognitive science researchers.

One reason for this is that design research requires significant domain facility and knowledge of the researcher. For example, if one wishes to study software design, it is necessary to have a pretty good understanding of how software and the software development process work—both technically and socio-technically. Many cognitive scientists are daunted or just unwilling to invest that level of preparation. As a result, cognitive science has focused on "universal" domains like human language, causal reasoning, and visual perception, or relatively accessible technical domains like K-12 education.

Accordingly, research on design occurs slightly out of the focus of the academic establishment of cognitive science, within the particular design domains themselves. Thus, it seems more common for an architect to learn about cognitive science and then to carry out investigations of architectural design than it is for a cognitive scientist to learn about architecture in order to pursue such investigations. Indeed, one finds the design literature scattered among domain-oriented journals and conferences, as well as in a couple interdisciplinary journals specifically addressed to design (*Design Studies* and *Design Issues*).

This eclectic institutionalization lends a special vitality to design research. Those who might be daunted by the need to grapple with domain-specific concepts, skills, and practices tend not to be among design researchers. But it turns out that many researchers are attracted by the special demands, and perhaps the special opportunities of studying design. Design research is a burgeoning area of cognitive science, albeit often outside the bounds of the establishment.

In *The Cognitive Artifacts of Designing*, Willemien Visser reviews and synthesizes a substantial body of empirical research and theoretical discussion about design—particularly the early stages of design—as cognitive activity. She seeks to reconcile two modern-classic lines of theorizing about design: One, following Herbert Simon among others, views design as problem-solving of ill-structured problems, the other, following Donald Schön among others, views design as improvised construction tightly coupled with reflection. Visser argues that the two views can be integrated in a concept of design as the *construction of representations*, where representations are taken to be artifacts, both cognitive and physical, that codify the current state of a design, while also suggesting directions for further development.

The task of this book is a critical task for the development of design research as cognitive science. Visser is summarizing and attempting to integrate a veritable mountain of empirical and theoretical work. Some of this work has definitely been carried out without much benefit of alternative views. There is a need to make comprehensive sense of the whole. Her work shows the possibility of moving from mere cumulation of concepts and results to a theory of design. I think that this book will help much to reintegrate design research and cognitive science to the benefit of both.

Visser herself has described some of the frontiers for pushing this agenda ahead: the role of emotion in design cognition—especially in the real workplace of design, design cognition in organizations, and later stages of the design process, including impacts of implementation on redesign. This book is a very significant contribution to design research, and more broadly to cognitive science. But I also look forward to these continuations of the research program. There is so much still to understand.

John M. Carroll

Preface

This book presents a new theoretical framework for cognitive design research: Designing as the construction of representations. Right from the outset, I feel I should formulate two words of warning with respect to these representations. (1) They may be external or internal (i.e., mental). (2) The term "representation" itself is somewhat misleading, due to its prefix "re-." The German and Dutch terms "Vorstellung" and "voorstelling" (Engl. "presentation") are less deceptive. When people construct a representation, they do not re-present an "original," a "reality" that "exists" beyond their experience, independently from their views on the world. It is through the representations they construct that people make present their views.

What is the link between design and representations? Designing consists in *specifying* an artifact, for example a machine tool—not in its *implementation*, its fabrication in the workshop. The result of design is a representation, the specifications of the machine tool. These representations are also artifacts, that is, entities created by people—they are "man-made as opposed to natural" (Simon, 1969/1999b). They are meaningful for people who elaborate them for specific uses (Stahl, 2002)—even if they may be used for other functions. Artifacts may be physical (machine tools, buildings, cars, or garments) or symbolic (software, social welfare policies, route plans, or any procedure); they may be internal (mental representations) or external (drawings, mock-ups). The term thus pertains not only to material objects. The antonym of an "artifact" is a "natural"—not an "immaterial"—entity.

Design results in artifacts. When ambiguity might arise, I use the expression "artifact product" for the artifact whose specification is the object of the design process and that remains to be implemented once the design process is over. This term allows it to be distinguished from the numerous intermediary artifacts and from the final artifact as constituted by the specifications.

A fundamental idea in this book is that artifacts are thus not only the outcome of design. In their representational form, they are constructed and used throughout the design process. When people use artifacts as tools in their cognitive activities (be they performed in an individual, autonomous or in a collaborative, interactional setting), these artifacts may be qualified as "cognitive." Such cognitive artifacts are not necessarily internal. Many external entities (whether physical or symbolic) function as cognitive artifacts. There are many forms of mental cognitive artifacts, such as rules of thumb, mnemonics, shopping lists, and all kinds of procedures. They are not discussed here. The mental cognitive artifacts of designing are often the more or less vague ideas or images that designers have "in their heads." Through externalization (in verbal descriptions, or in diagrammatic representations, such as sketches and other drawings) and interaction with colleagues, these mental cognitive artifacts continue their evolution towards the final artifacts, which, by definition, are external representations: People other than the designers are to implement these specifications into the artifact product.

External, material cognitive artifacts are the most frequently analyzed in the literature: They range from marble ashtrays used as status symbols or as hammers, via maps and calendars, to computer systems. A prototypical example of external, material cognitive artifacts used in design are diagrammatic representations and mock-ups of what designers envision as the artifact product to be designed.

Cognitive artifacts can be examined from many angles. Their development in the young child, their mediating role and function as tools, have been analyzed by Vygotsky and followers. Norman (1991) focuses on physical cognitive artifacts. I focus particularly on the construction of both internal and external cognitive artifacts in the representational activities of designing, and on their mediating role—as a tool both between a designer and her- or himself, and between the designer and her or his colleagues.

The book is composed of four parts with an introduction (Part I) and a conclusion (Part VI). Parts II, III, and IV discuss the main theoretical viewpoints adopted in cognitive design research. Against this background, Part V presents my own viewpoint on design.

Part II (chapters 7–9) presents the classical view on design, embodied by Simon's symbolic information processing approach (SIP), which analyses design as problem solving. Simon's *Sciences of the Artificial* has been one of *the* references in design studies—and this holds for all domains, both engineering and cognitively oriented approaches. It has nevertheless been the object of much debate—as has been the entire "cognitivistic" approach to human activity. Part III (chapters 10–13) presents modern views on design that have questioned the SIP paradigm. They are represented by the situativity approach (SIT), especially in the form that Donald Schön has given to this framework in his reflective-practice approach. Part IV (chapters 14 and 15) confronts the classical and modern views on design. My own view on design, presented in Part V (chapters 16–22), is to consider that design is most appropriately characterized as a construction of representations. The separation that I introduce in my presentation, between the representational structures that are built up and the corresponding representation-construction activities, is of course a purely analytical one, and in the conclusion of Part V, I propose some links between the two. Another division is that between representations according to the stage in the design process in which they take place. Therefore, I dedicate three consecutive chapters to representations at the source of a design project (the requirements, which, from a problem-solving perspective, are the "design problems"), intermediate representations, and representations at the end of a design project (the specifications, or "design solutions"). In the *Discussion* chapter of Part V, I state that, while adhering to the so-called "generic design" position—that is, design is a specific cognitive activity, distinct from other types of cognitive activity—, I also consider that designing may take different forms depending on the nature of the artifact. I put forward a series of candidates for dimensions to distinguish between these different forms of design. In the Concluding part (Part VI, chapters 23–25), I establish a link between SIP- and SIT-inspired research, on the one hand, and studies into individual and collective design, on the other. I conclude my discussion of the relations between design and problem solving, defending the position that even if design *involves* problem solving, it *is* more than problem solving. I conclude my book with the claim that my "design as a construction of representations" proposal opens up new possibilities for the study of design, providing new views on designers' needs for assistance and the potential corresponding support modalities.

Design has many facets. Long as this book is, many aspects of design are covered only in passing—or even not touched upon. I adopt a cognitively and theoretically oriented viewpoint on design, grounded in empirical data from mostly observational studies conducted on design professionals working on industrial design projects. I address only indirectly how these data and my theoretical proposals might be applied, for example, in design education and assistance. Design methodology is discussed, but merely from the viewpoint of the underlying prescriptive, generally stage models, in a confrontation with the descriptive, generally process models on which I focus.

My primary audience is researchers and students in the domain of design, not only in cognitive design studies, but also design methodology and engineering. Yet, this book is also relevant for cognitive psychologists and researchers from other cognitive sciences who are interested in problem solving in the "real world." This holds, in particular, for psychologists concerned with a critical view on the classical models of problem solving (notably, Simon's symbolic information processing framework). In addition, my analysis of the various activities involved in the construction of representations provides new perspectives on this classical theme in cognitive psychology. For the many cognitive ergonomists involved in design situations, this book offers a theoretical foundation for their interventions. Design practitioners who wish to find immediately applicable ideas for their practice may consider that the proposals advanced in this book do not match their everyday needs and that their practice cannot directly benefit from the considerations formulated. However, reflection and externalization, through verbalization and discussion, concerning one's activity may be a powerful tool for cognitive development with respect to the object of reflection—here, designing. It generally leads to restructuring one's representations and one's knowledge—and thus transforms one's activity.

Acknowledgments

I wish to thank a large number of people:

- The members—present and past; research scientists, students, and administrative assistants—of the EIFFEL2 team and its preceding INRIA research groups, especially Françoise Détienne, Pierre Falzon, and André Bisseret, who have been not only my bosses, but also my friends. Special thanks to Thierry Février Quesada, who proved such calm and cheerful company and good support over this last year.

- Jacques Perrin, whose invitation to represent the cognitive-ergonomics view on design, at the 2002 International Conference in honour of Herbert Simon "The Sciences of Design. The Scientific Challenge for the 21st Century" triggered a paper in which I began to put together into an integrated view, the thoughts and analyses of design I had been developing for some 20 years.

- John Carroll, whose suggestion to submit this paper to *Human–Computer Interaction* led to a text that was much too long—and subsequently became this book.

- John Gero, who two days after a research report presented this work in progress on internet, invited me to come to Australia, to talk about the view on design proposed.

- All the designers whom I "observed," following them in their daily practice, and whose indirect contribution formed the basis of my research.

- Six anonymous reviewers who, four at HCI and two at LEA, formulated remarks and questions that made me reflect even more on the subject matter of this book.

- Finally Richard James who so often reviewed my English papers—I did not dare to ask him to go through this entire book, but I hope I learned enough for the present text not to contain too many errors.

Rocquencourt, May 31, 2006

Willemien Visser

I

Introduction

Design is an important, all-pervading domain of human activity. Not only are new sport cars and mobile phones the object of design, but so too are artifacts as diverse as traffic signals (Bisseret, Figeac-Letang, & Falzon, 1988), meals (Byrne, 1977), route plans (Chalmé, Visser, & Denis, 2004; B. Hayes-Roth & Hayes-Roth, 1979), software (Détienne, 2002a), and, of course, buildings of all kinds (Ö. Akin, 1986b; De Vries, 1994; Hamel, 1989; Lawson, 1994; Lebahar, 1983; Schön, 1983).

This book adopts a *cognitive* viewpoint on design; namely, it focuses on the types of cognitive processes and structures implemented by designers. Analyzed from this viewpoint, design projects involve a multifaceted activity, which is of interest, first, to cognitive psychology and cognitive ergonomics, but also to disciplines that examine design from other viewpoints, such as design methodology, artificial intelligence (AI), and engineering-design disciplines. As developed thoroughly in what follows, we consider that design is most appropriately characterized as a *construction of representations*.

"Cognitive design studies" is the qualification that we have come to use for studies of design focusing on its cognitive aspects (Visser, 2002c, 2006). In the engineering domain, researchers often qualify their empirical studies as research on "design thinking" (Buchanan, 1992; Cross, Dorst, & Roozenburg, 1992; Dörner, 1999; Galle, 1996; Rowe, 1987), sometimes distinguishing "design reasoning" and "design intuition" (Ö. Akin, 1992, p. 39).

Cognitive design studies have been developed since the 1960s, but their results still often seem to be acknowledged only in rather restricted circles. In the domain of human–computer interaction (HCI), for example, their relevance is of course recognized by many researchers, but most authors do not show great familiarity with the existing tradition and the research that has already been realized in this field. Yet, HCI is undeniably a field with close relations to cognitive psychology and cognitive ergonomics, with design as a central area of both research and development.

In general, however, the importance of this type of research is increasingly realized. The American National Science Foundation (NSF), for example, formulated, at the beginning of 2004, a Program Solicitation under the title "Science of Design" focusing on the design of "software-intensive systems" (*Science of Design*, 2004). The solicitation aims "to stimulate research and education projects that build the Science of Design." "Studies of designs, designers, and design methodologies" are among the topics that are quoted as relevant to this solicitation.

Our aim in this book is to present materials for a *cognitive descriptive model of design* that, on the one hand, furthers our understanding of design and, on the other hand, offers elements to people aiming to advance the education and the practice of professional designers.

Outline of this part. We first sketch the organization of this book. Then, after a very short presentation of the aspects of design on which the books focuses and of the public it addresses, we give some historical pointers. The next chapter presents two series of preliminary terminological issues, namely definitions in the domains of problem solving and data collection. A chapter discussing the different types of models currently developed and used in design research and design practice is followed by the closing chapter of this introductory part, which presents our empirical design studies at a glance.

Organization of the book. This book is organized in the following way. After this introductory Part I, Part II describes the classical cognitive viewpoint on design, namely the symbolic information processing (SIP) approach, represented by Herbert A. Simon. Part III focuses on the main alternative to the SIP approach proposed for design, namely the situativity (SIT) framework, which Donald Schön applied on design in the form of his reflective-practice approach. Part IV compares the SIP and SIT approaches to design. Part V is the main part of this book. It describes the approach to design that we have been developing for some 20 years now. It presents nuances and critiques with respect to both SIP and SIT approaches, and completes and integrates these two approaches into our own cognitively oriented dynamic approach to design. In Part VI, the *Conclusion*, we review and discuss the different approaches to design presented in the book. We conclude by a chapter on the proposal defended herein—design as a construction of representations. We sketch some directions for further research and relate our proposal to new views on designers' needs for assistance and the potential corresponding support modalities.

1 Focus of This Book

This book focuses on cognitive, dynamic aspects of real design, that is, the actual cognitive activity implemented by designers during their work on professional design projects. It is hoped that the elements presented will be used to build a descriptive cognitive model of design. We address only indirectly their possible applications, such as in design education (Adams, Turns, & Atman, 2003; Demirbas & Demirkan, 2003; Expertise in Design, 2004; A. Heylighen & Verstijnen, 2003; Oxman, 2004) and assistance (Atwood et al., 1995; Bertolotti, Macrì, & Tagliaventi, 2004; Falzon & Visser, 1989; Fischer, Nakakoji, Ostwald, Stahl, & Sumner, 1998; Marmaras & Pavard, 1999; Visser & Hoc, 1990).

This chapter makes explicit a number of presuppositions underlying the aforementioned qualifications. They are familiar to researchers from cognitive psychology and ergonomics, but may be informative for readers from other disciplines. A special chapter (chap. 4) dedicated to terminological issues details related concepts that are central to the discussion in this book.

The term *cognitive aspects* refers to designers' *cognitive processes and structures (knowledge and representations)* as distinguished, especially, from sociocultural and emotional aspects of design, which are peripheral to the discussion in this book. In spite of its essential role in design, especially of physical artifacts, perception is discussed only *in passim*. Issues related to environmental and ecological aspects of design (ecodesign or sustainable design) are not covered either (for references, see, e.g., Méhier, 2005).

Dynamic refers, in the framework of this book, to the *generation* and *use* of knowledge and representations, rather than to their *content*.

Cognitive activity refers to *the way in which people realize their task at a cognitive level*.

The focus on "real design" points toward *design as performed in a designer's usual working situation*—rather than in artificially restricted conditions, such as laboratory experiments. We focus on *actual* design rather than the design task or the design process that are the reference in design methods, and in prescriptive and stage models. In the rest of this book, *design activity* or *designing* refers to this real, actual cognitive design activity. Terms such as *steps*, *stages*, *actions*, *methods*, and *tasks* may refer to theoretical, abstract entities not necessarily observed—or even observable—in real design projects.

With respect to the artifacts products of design that are central in this book, we focus on design of *products* (material and symbolic) in its opposition to *production systems* (for design of production systems from an ergonomic-intervention viewpoint, see Daniellou, 1999; Desnoyers & Daniellou, 1989).

1.1.   ELEMENTS FOR A COGNITIVE
DESCRIPTIVE MODEL OF DESIGN

This book presents components for a cognitive descriptive model of design—
neither a prescriptive, nor a stage model. It does not pretend to completely specify such a model. Some components are going to be more detailed than other
ones. In our own studies, for example, the organizational and reuse components
of design have been especially analyzed and described in detail. Other components require that the already available data be completed and formalized, and
still other parts even require that research be initiated.

1.2.   EMPIRICAL BASIS

Most data in this book comes from empirical studies. Much material underlying
the presentations is based on our own work, that is, mainly empirical cognitive
design studies conducted, over some 20 years, on professional designers working in different task domains (software, mechanical, and industrial design; see
Table 6.1). These are mainly field studies, but we have also conducted experimental research. Of course, the results of other empirical cognitive design studies, conducted by colleagues, have also been integrated in our presentations.

    In several studies, we have shown that the analysis of real activities may lead
to qualification and even modification of the theoretical frameworks and conclusions that are formulated based on experimental research performed in controlled, generally artificially restricted situations. In our study on categorization
and types of expertise (Visser & Falzon, 1992/1993), for example, we qualified
the classical theoretical framework in cognitive psychology regarding categorization: we introduced a distinction between "operative" and "task-dependent"
categorization. In our software-design study (Visser, 1987b), we showed how a
professional designer in an industrial design project uses different strategies than
previously observed in laboratory studies of "programmers."

1.3.   FOCUS ON THE EARLY STAGES
OF DESIGN

Our focus is largely on what prescriptive models consider the "early stages" of
design ("preliminary," "logical," or "conceptual" design).

    Economic analyses of design demonstrate the importance of these early
stages. The study of alternatives is much less expensive, for example, if it is
conducted near the beginning of a design project (Perrin, 1997, quoted in Perrin,

1999). Several authors advance percentages concerning the corresponding cost and possible gain. They affirm, for example, that the first 5% of the design process commits more than 75% of the overall costs to market the product (Carrubba, 1993, quoted in Hsu & Liu, 2000; Sharpe, 1995). To be able to examine various alternatives, to confront the perspectives of the different stakeholders in the design project, and to attain pondered compromises, more time should be allocated to these early stages. The resulting supplement in duration, and thus in cost, would be largely compensated afterward. This is, therefore, one of the justifications advanced for the concurrent-engineering approach to design (Perrin, 1999).

1.4.   INDIVIDUAL AND COLLECTIVE DESIGN

Because this section uses two notions that may require definition—distributed design and co-design—we start with a terminological paragraph.

Distributed design and co-design. In distributed-design situations, several designers are involved simultaneously, but not together, in the same collective process. They carry out well-determined tasks that have been allocated beforehand to them, and they pursue goals, or at least subgoals, that are specific to them. One of their objectives is to participate as efficiently as possible in the collaborative resolution of the problem.

In co-design situations, design partners develop the solution together: They share an identical goal and contribute to reach it through their specific skills. They do so with strong constraints of direct cooperation in order to guarantee the success of the problem resolution.

Studies focusing on individual design are sometimes considered artificial, or even irrelevant: Design always involves several people—at least two, a client and a designer, or a designer and a user. Given that design projects generally require the integration of information and knowledge from a variety of domains, they usually involve multiple competencies, and thus collaboration between various design participants from different areas of expertise. In our view, however, data concerning individual design is particularly relevant, and today's focus on collective design may even obscure the importance of individual design activity (see also Stacey & Eckert, 2003, p. 155).

First, even if many design projects are undertaken by large teams involving big numbers of so-called "designers" and other participants, and even if discussion, negotiation, and synchronization (examples of collaborative-design activities) play a crucial role in the elaboration and selection of solutions, an important proportion of design activity remains the work of single individuals, especially during distributed design (for software engineering, see, e.g., Kitchenham

& Carn, 1990). In their article titled "Resume of 12 Years Interdisciplinary Empirical Studies of Engineering Design in Germany," Pahl, Badke-Schaub, and Frankenberger (1999) advance that "in design processes individual work dominates to an extent of > 70%, compared to < 30% of teamwork" (p. 490). These conclusions are based on protocols collected during 28 weeks of observational studies in the industry (E. Frankenberger, personal communication, January 17, 2001).

In addition, even during co-design, cognitive activities are those implemented in individual design augmented by other activities that are specific to cooperation (especially coordination, operative synchronization, construction of inter-designer compatible representations, conflict resolution, management of different representations through confrontation, articulation, and integration—all these largely via argumentative activities). As we have defended elsewhere (Visser, 1993a), we do not see any evidence to suppose that cooperation modifies the nature of the basic activities and operations implemented in design (i.e., generation, transformation, and evaluation of representations).

Finally, the development of appropriate work environments, such as shared and private work spaces in (computer) mediated design, requires the analysis of the articulation between the different forms of reasoning implemented in both individually and collectively conducted activities.

On the other hand, collective working is not the panacea for all multifaceted processes—contrary to what may be suggested by affirmations such as: "Given design's complexity, it has to be conducted in a collective setting." In certain conditions, individually conducted activities may be more appropriate. Creativity—often considered essential in design projects—may, for example, require individual activity in certain situations, as has been shown by Van der Lugt (2000) for idea reinterpretation in creative problem-solving meetings.

1.5.   COGNITION AND EMOTION

Given that research on the role of emotion in the domain of design cognition is just in its early stages, we review, in this section, the few elements that the literature provides.

Cognition is frequently opposed to emotion, as referring to "affective" states, events, and experiences. The history of psychology confirms this view: The classical, main theories of cognition (Anderson, 1983, 1993; Newell, 1990) do not take into account emotion. Recently there are, however, authors who have started to integrate these two faces of human beings (Cacioppo & Gardner, 1999; Carlson, 1997; Ribert-Van De Weerdt, 2003). There is indeed "a great deal of evidence that emotion affects cognition in a variety of ways" (Carlson, 1997, p. 112)—and that cognition affects emotion, as shown by Ribert-Van De Weerdt. This double relationship leads Ribert-Van De Weerdt to conclude that emotions are an integral part of the cognitive system, even if the nature of this relation is not yet clear. Discussing emotion in the presentation of his experienced-cognition framework, Carlson formulates the same conclusion, specifying an element of relation between the two, namely information. "Recognizing the informational, but not necessarily content-constituting, contribution of emotional arousal to experienced mental states lets us see that emotion and cognition are not separate domains but theoretically separable aspects of experienced cognition" (p. 124) (see also S. Moore & Oaksford, 2002, who present examples of computational models of emotion and cognition–emotion interaction; see as well the journal *Cognition and Emotion*).

Although we adhere entirely to this position, this book hardly ever discusses emotional aspects of design—first because of the lack of relevant research. Second, even if emotions have been studied in psychology for many years already, be it independently of cognition, they have rarely been studied in the workplace (see Ribert-Van De Weerdt, 2003, for a review of research on this issue that has been developed lately in psycho-ergonomics). Ergonomics traditionally distinguishes physical and cognitive aspects of tasks, analyzing the contribution of each in human activity, but the contribution of emotions is still an ignored dimension. In a professional context, emotions nevertheless play a part in various ways. The interaction of emotion and cognition, in general, has "*content* effects such as the apparent mood-specificity of memory, and *noncontent* effects such as the disruption of cognitive (or other) performance by extreme emotional arousal" (Carlson, 1997, p. 123). In their work, people may have to cope with "heavy demands relative to these psycho-affective dimensions" and "the number of work situations [concerned] is increasing" (Ribert-Van De Weerdt, 2003, p. 9). However, emotion has not only disruptive effects. These effects "are perhaps best documented . . . but Damasio (1994) and others pointed out the importance of emotionality for the normal control of activity" (Carlson, 1997, p. 119). The so-called "Yerkes–Dodson law suggests that for a particular task performed by an individual, there will be an optimal level of arousal" (Carlson, 1997, p. 124). This may hold for all types of professionals, for designers as well as for stockbrokers.

Emotion is thus involved in the control of activity. "In decision making, the emotional consequences of how choices are framed may affect which choices are preferred (A. Tversky & Kahneman, 1981). These phenomena suggest that the arousal associated with emotional experience sometimes contributes to the degrees of commitment associated with beliefs" (Carlson, 1997, p. 120). One may imagine the effects of this factor on all activity that is performed in a collective setting, such as many design projects.

In addition to these effects, the emotional impact of the object one is working on may be a dimension that people might profitably take into account. In the case of design, this is the domain of "emotion-driven design." Certain designers take seriously the emotions of their artifacts' users, aiming to design artifacts that these users would experience as pleasant or desirable. Desmet and Dijkhuis (2003) present an application of this design approach as they have used it to design a children's wheelchair (see also Desmet, 2002). At first, the emotional responses to existing products (wheelchairs, in the example) are assessed. The authors use a nonverbal self-report instrument. The results of the assessment are then transformed into starting points for a new design. Discussing the difference that exists between "understanding how products evoke emotions, and actually being able to manipulate the emotional impact of a design" (pp. 26–27), the authors refer to theoretical models of emotions (associated with products). The authors propose a completely new approach to product design and development, referring to work with titles such as "Pleasure with products: Beyond usability," "Funology: From usability to enjoyment," "Designing products with added emotional value," and "Design and emotion."

A specific proposition concerning the role of affective processing (based on "feelings," which he distinguishes from "emotions") in design cognition is formulated by Love (1999). The author presents two main ways in which feelings and "feeling-based affect" may play a role in design. "Closure is [one way in which] affect impacts on human design cognition—the other is the way that affect influences which new thoughts are initiated in response to a situation." In the context of the paper, "closure is that human activity that is involved in deciding, in broad terms, whether to initiate a process, continue a process, or to stop a process." Through feeling-based closure, affective processes occur, passively, in testing (or optimizing) design solutions: Designers generate a solution to a design problem and then close the process if they "[*feel*] satisfied with the solution" (or "[*feel*] that it is better than the previous best solution") and are "happy to *close* the process." Affect plays an active part in design cognition through its role in solution generation, through its influence on "*thinking* about a problem." Love presents his approach as a "new basis for computerising aspects of nonrational design cognition" and outlines "how this computerisation of affectively-based processes might proceed in order to provide better assistance for designers." The author refers to a number of designers and design researchers who mention the role of feelings in designing (see also his references to different research projects and applications by the Massachusetts Institute of Technology—Affective Computing Research Area).

We close this section by mentioning that there is already, at least, one specialized international network in the domain, with its scientific events. The Design & Emotion society aims to "integrate salient themes of emotional experience into the design profession." The society was established in 1999 as "an international network of researchers, designers, and companies sharing an interest in experience driven design." It organizes the biannual International Conference on Design & Emotion as well as Design & Emotion workshops (the fifth International Conference on Design & Emotion is held in 2006; retrieved May 17, 2006, from http://www.designandemotion.org/de11.php).

Readers may notice that we do not cover at all the fast-growing amount of research in AI and HCI concerned with emotion. There is, for example, the Network of Excellence HUMAINE, Human-Machine Interaction Network on Emotions (launched on January 1, 2004, for 4 years, in the framework of the IST FP6 Multimodal Interfaces thematic priority), which aims at "emotion-oriented systems."

2 Public Addressed

The nature of this book is theoretical—but it is entirely based on empirical data from mostly observational studies conducted on design professionals working on industrial design projects. It mainly addresses researchers and students in the domain of design, not only cognitive design studies, but also design methodology and engineering.

For at least two reasons, this book may be relevant for cognitive psychologists. First, our critical analysis of Simon's positions concerning problem solving, as applied to design cognition, may introduce some fresh air into certain psychology laboratories. Second, cognitive psychologists may find interesting the analysis of problem-solving and representation-construction activities in real-world situations, rather than in a laboratory context.

Cognitive ergonomists who proceed to interventions in design situations may find in this book a theoretical foundation for their activity as well.

The situation is somewhat different for design practitioners. The way in which their practice may take advantage of the data presented here is rather indirectly, namely through support systems (tools, environments, and education) whose development is based on the structures and processes identified in cognitive design studies. Design practitioners who wish to find immediately applicable ideas for their practice may consider that the proposals advanced in this book are far from their experience and that their practice cannot directly benefit from the considerations formulated in this book. The cognitive analysis of complex activities such as design may reveal cognitive structures and processes that people themselves not only are often unaware of, but also do not necessarily recognize as relevant. It is not by becoming aware of them that they will necessarily modify their activity. One may notice, however, that reflection and externalization (through verbalization and discussion) concerning one's activity may be a powerful tool for cognitive development (Mollo & Falzon, 2004), generally restructuring one's representations and one's knowledge—and thus transforming one's activity.

These reflective functions are observed in the context of metafunctional activities. As defined by Falzon et al. (1997, quoted in Mollo & Falzon, 2004), a "metafunctional activity" is a reflective activity by which people take their work as an object of reflection. It may give rise to the elaboration of internal (cognitive) or external tools, for possible future use. It may thus lead to improvement of existing tools and of people's knowledge on these tools, and to elaboration of new tools (as claimed by Falzon et al., 1996, 1997, quoted in Mollo & Falzon, 2004). Reflective activity therefore plays a central role in the construction and evolution of technical knowledge.

This analysis can be related to ideas developed by Carroll (2006) in his elaboration of Simon's "Science of Design" with respect to "dimensions of participation." Participatory design may lead to human development. *Doing design* together "can provide not merely a common topic, a shared orientation to knowledge, but an activity for engaging knowledge that makes learning and human development ineluctable" (Carroll, 2006, p. 15). In a long-term design project with a group of public school teachers (see Carroll, Rosson, Chin, & Koenemann, 1997, 1998), Carroll and his colleagues observed "the defining characteristics of developmental change" at work. Referring to Piaget and Inhelder (1969), and Vygotsky (1978), Carroll (2006) qualified these characteristics as "active resolution of manifest conflicts in one's activity, taking more responsibility, and assuming greater scope of action" (p. 16).

As formulated clearly by K. Friedman (2003), "design theory is not identical with the tacit knowledge of design practice" (p. 519). Of course, such knowledge is vital, and not only to practice. Theory, however, is based on knowledge that is not only made explicit, but that is also more general than the idiosyncratic knowledge, know-how, or other experience-based expertise of a particular practitioner. Theory requires abstraction from specific, personal knowledge. K. Friedman especially insists on the "explicit articulation" of knowledge for theory formation, because otherwise there is no possibility for communication and joint reflection on each one's representations, and for contrasting and testing them. This is why he considers as a misguided attempt, which inevitably leads to a dead end, the effort "to link the reflective practice of design to design knowledge, and . . . to propose tacit knowledge or direct making as a method of theory construction" (p. 520).

3 Some Historical Pointers

Bayazit (2004) draws a detailed picture of the history of design research, starting in the early 20th century. He first presents precursor movements (such as De Stijl and Bauhaus) and then authors with social psychology, ergonomics, and cultural-anthropology orientations (such as Broadbent and colleagues) and engineering-oriented authors (such as Hubka, Eder, Asimov, Ullman, Archer). He closes with a discussion of artificial-intelligence and other cognitive-science researchers.

Until some 10 to 15 years ago, most design research was concerned with individual design. In more recent years, an important shift in research on the cognitive aspects of design has consisted in starting to study collectively conducted design activities—from then onward, perhaps even to a greater degree than individual design (Blessing, Brassac, Darses, & Visser, 2000; Détienne, 2002b; Falzon, Darses, & Béguin, 1996; Lindemann, 2003). Yet, there are also some older studies on cooperation in design projects (see, e.g., M. Klein & Lu, 1989; Krasner, Curtis, & Iscoe, 1987; Walz, Elam, Krasner, & Curtis, 1987).

A few studies establish a relationship between individual and collective design (Visser, 1993a, 2002a). The 1994 Delft workshop (Cross, Christiaans, & Dorst, 1996; Dorst, 1995) offered the occasion to compare how an individual and a team solved the same design problem. The organizers indeed provided a group of design researchers with two videos (plus protocol transcriptions) corresponding to a design of a mountain bike attachment as executed by both an individual designer and a team of three designers. Most researchers analyzed only one of the videos and protocols, but some proceeded to a comparative analysis (Dwarakanath & Blessing, 1996; Günther, Frankenberger, & Auer, 1996).

The change in focus from individual to collective—which has also been observed in other task domains—has been accompanied by the acceptance of design as a central field of research in cognitive ergonomics for which specific analysis methods are to be developed (Darses, Détienne, Falzon, & Visser, 2001; Visser, Darses, & Détienne, 2004). Nevertheless, a great number of studies of design are still conducted in artificially restricted laboratory conditions, in which the design situation is rather different from that in professional design situations. Compared, however, to other activities analyzed as "problem solving," design has started to be examined rather frequently in actual working situations. After a few older studies on collective design in a professional setting (see, e.g., Krasner et al., 1987; Visser, 1993a; Walz et al., 1987), recently this form of design is even being studied in large-scale industrial settings (Darses, 2002; Détienne & Falzon, 2001; Martin, Détienne, & Lavigne, 2001).

Cognitive design research started long after other studies on problem solving. There are some rare precursor studies in the 1960s and 1970s of the 20th century (Ö. Akin, 1979/1984; Byrne, 1977; Darke, 1979/1984; Eastman, 1969, 1970; B. Hayes-Roth & Hayes-Roth, 1979; Lawson, 1979/1984; Marples, 1961; Reitman,

1964, 1965; Thomas & Carroll, 1979/1984; see Robert, 1979, for references to other early empirical design studies).

Except for these few pioneering studies, it is only since the 1980s that empirical work in this domain has started (Adelson, 1984; Ö. Akin, 1986b; Baykan, 1996; Bucciarelli, 1984; Carroll, Thomas, & Malhotra, 1980; Chatel & Détienne, 1996; Darses, 1994; Détienne, 1991a; Guindon, Krasner, & Curtis, 1987; Hoc, 1987; Lebahar, 1983; Malhotra, Thomas, Carroll, & Miller, 1980; Michard, 1982; Rist, 1990; Rosson & Alpert, 1990; Schön, 1984; Suwa & Tversky, 1997; Ullman, Staufer, & Dietterich, 1987; Visser, 1987b; see also Fig. 1 in Cross, 2001a, p. 80).

The first empirical studies of design concerned *architecture* (Ö. Akin, 1979/1984; Eastman, 1969, 1970; Foz, 1973; Krauss & Myer, 1970; Lawson, 1979/1984; Traverso & Visser, 2003; see other references in Cross, 1984a; De Vries, 1994; Hamel, 1989) and *mechanical and industrial engineering* (Ullman, Dietterich, & Staufer, 1988; Visser, 1994a; Whitefield, 1986). Subsequently, *software* design has become the object of much cognitive design research (see our later subdivision). Today, *engineering* design is one of the central domains in which design is being examined, often in collective settings (Ball, Evans, & Dennis, 1994; Frankenberger & Badke-Schaub, 1999; Kavakli, Scrivener, & Ball, 1998; Martin et al., 2001; McGown, Green, & Rodgers, 1998).

Other domains in which empirical studies have examined design from a cognitive point of view are *musical composition* (Reitman, 1964, 1965), *computational geometry algorithm* design (Kant, 1985; Kant & Newell, 1984), *traffic-signal setting* (Bisseret et al., 1988), *computer-network* design (Darses, 1990b, 1990c), *database conceptual modeling* (Karsenty, 1991b), *industrial* or *product* design (Cross, 1997; Cross et al., 1996; Visser, 1995b; see also Archer, 1965/1984; Dorst, 1997; Valkenburg & Dorst, 1998), and *composite-structure* design (Bonnardel, 1991a; Visser, 1991b, 1993a).

Less common activities that have been analyzed as design are certain *letter-writing and naming activities*, or even *communication* in general (Thomas & Carroll, 1979/1984), *knowledge modeling* (Visser, 1993b), *errand planning* (Chalmé et al., 2004; B. Hayes-Roth & Hayes-Roth, 1979), *text composition* (Bisseret, 1990; Hayes & Flower, 1980), *meal planning* (Byrne, 1977), and *space planning* or *interior* design (Carroll, Thomas, & Malhotra, 1980; Eckersley, 1988; see also references in Bayazit, 2004; Cross, 2001a, 2004b; Eastman, 2001).

Software "design" studies. Research on design in the domain of software engineering occupies a special position. Many studies on "software design" rather concern coding.

From the 1980s on, cognitive design studies have been conducted in the software-engineering domain (Carroll, Thomas, & Malhotra, 1980; Carroll, Thomas, Miller, & Friedman, 1980; D'Astous, Détienne, Visser, & Robillard, 2004; D'Astous, Robillard, Détienne, & Visser, 2001; Davies, 1991b; Détienne, 1986, 1990, 1991b; Gilmore, 1990a; Gilmore, Winder, & Détienne, 1994; T. R.

G. Green, 1980; Guindon, 1990a, 1990b; Hoc, Green, Samurçay, & Gilmore, 1990; Jeffries, Turner, Polson, & Atwood, 1981; Malhotra et al., 1980; G. M. Olson, Olson, Carter, & Storrøsten, 1992; Ormerod, 1990; Pennington & Grabowski, 1990; Rist, 1991a, 1991b; Thomas & Carroll, 1979/1984; Visser, 1987b). Several paradigms have been examined. The first studies concerned procedural languages such as Fortran and Pascal (R. E. Brooks, 1977). More recently, studies have been examining the object-oriented paradigm (Détienne & Rist, 1995; Herbsleb et al., 1995), mostly through studies of C++ (Détienne, 2002a). LISP (Weber, 1991) and a declarative Boolean language have also been studied (Visser, 1988b).

For at least two reasons, several of these studies do not confer much knowledge concerning design, however. First, many of them concern programming in the sense of coding, that is, implementation of design decisions, rather than design, that is, the elaboration of these decisions. Even if in several of these studies on coding, the participants also had design tasks to fulfill, few authors have analyzed the corresponding design activities. This may explain why, in its early years, the research domain was called "psychology of programming," and the workshop that has come into life in these years was entitled "Empirical Studies of Programmers" (ESP) (Cook, Scholtz, & Spohrer, 1993; Gray & Boehm-Davis, 1996; Koenemann-Belliveau, Moher, & Robertson, 1991; G. M. Olson, Sheppard, & Soloway, 1987; Soloway & Iyengar, 1986; Wiedenbeck & Scholtz, 1997; see also Gilmore et al., 1994; Hoc et al., 1990).

Second, most of these studies concern small tasks—be it design or coding—examined in artificially restricted laboratory conditions, whereas many distinctive characteristics of design appear only in real, professional design (Visser, 1987b).

Our longitudinal study concerning the design of industrial programmable-controller (IPC) software, conducted in the years 1987–1988 (Visser, 1987b; see Table 6.1), was among the early "programming in the large" research—as opposed to the "programming in the small" research that characterized most studies conducted in those years. One of the questions raised at the end of the First ESP workshop (Soloway & Iyengar, 1986) and which subsequently became the title of a Future Directions paper, was "By the way, did anyone study any real programmers?" (Curtis, 1986). In this IPC study, we indeed analyzed the design activity of a "real programmer" and corroborated the hypothesis at the source of the study that strategies used in a professional design setting differed, at least partially, from those that had been observed until that day in studies conducted on most novice, student programmers working on artificially limited tasks.

Recent software-design studies focus on teamwork. The collaboration examined generally concerns designers, not designers and users or clients (but see Karsenty, 1991a).

4 Preliminary Terminological Issues

In order to reduce sources of misunderstanding between researchers from different disciplines, we include this chapter dedicated to terminological issues in the domains of "problem solving" and "data collection." Those familiar with cognitive psychology or ergonomics may skip the chapter.

4.1. DEFINITIONS IN THE DOMAIN OF PROBLEM SOLVING

Much misunderstanding and disagreement seem to be due to the specific, technical sense of "problem" in cognitive psychology (something also noticed by Stacey & Eckert, 2003, Note 3, p. 179). As is developed in this section, in cognitive psychology, "problem" qualifies one's task according to the representation that one has constructed of this task.

Here we discuss briefly the notions "problem" and "problem solving," which, given certain remarks and discussions in the cognitive design literature, seem to require comment.

"Problem" is a term that one encounters of course frequently in a prescientific sense—also in scientific publications. We expect that generally it will be obvious if this text uses the term with such a prescientific acceptation, referring, for example, to a difficulty, a challenge, or a dilemma. The same holds for "problematic." When Schön (1983) writes, for example, that a designer, "because he finds the situation problematic, . . . must reframe it" (p. 129), we take "problematic" as a nontechnical term.

It may also be noticed that the term "problem solving" often implicitly refers to the SIP paradigm, that is, the approach to problem solving developed by Newell and Simon in their famous *Human Problem Solving* (1972; for a succinct presentation, see Simon, 1979). An example is the following. Schön (1983) writes that "from the perspective of Technical Rationality, professional practice is a process of problem solving. Problems of choice or decision are solved through the selection, from available means, of the one best suited to established ends" (p. 40). In this context, Schön seems to use the term "problem solving" in a technical sense, to be precise, the one adopted by Newell and Simon, who are, for many SIT-inspired researchers, the proponents of "Technical Rationality." However, Schön continues: "But with this emphasis on problem solving, we ignore problem setting, the process by which we define the decision to be made, the ends to be achieved, the means which may be chosen. In real-world practice, problems do not present themselves to the practitioner as givens. They must be constructed from the materials of problem situations which are puzzling, troubling, and uncertain" (p. 40). When Schön introduces "problem setting" beside

14

"problem solving," he seems no longer to reason in terms of the SIP framework and seems to introduce—implicitly—a different approach. Yet, this approach still includes a component activity "problem solving." The scope of this "problem solving" must be different than in the SIP framework. Yet, Schön does not define it.

Designing and Problem Solving

The relation between design and problem solving is a complicated one. For a long time, inspired by Simon's (1969/1999b, 1973/1984) analysis, the position in cognitive design research was that design is problem solving. Opposition to this view has been formulated from the 1970s on (Rittel, 1972/1984; Rittel & Webber, 1973/1984), but only in the 1980s, especially due to people such as Schön (1983, 1988, 1992), new paradigms to design research have gained power and a large following. We discuss these alternative paradigms in Part III.

As this book aims to advocate, we consider that design *involves* problem solving, but that design *is* not problem solving: It is *not only* and *not mainly* problem solving. From a formal point of view (based on the technical cognitive-psychology definitions of "problem" and "problem solving"; see later discussion), one may classify design as "problem solving." We consider, however, that such a qualification does no justice to the nature of design.

Another argument that may be invoked in support of the position that design *is* not problem solving is the importance of sociocultural factors in design (which interact with its cognitive aspects).

A "Problem" is not a "Difficulty" or a "Deficiency"

It may seem strange to start a discussion of a concept by a presentation of what the concept *is not*. However, given the presuppositions—sometimes made explicit—in relation to "problems" and "problem solving" that underlie many papers about design, we consider it best to first discuss these presuppositions.

In many prescientific texts and contexts, but even in research papers and other scientific publications, the term "problem" is used as a synonym of "difficulty" or "deficiency." It is also often associated with the idea that a "problem" is being solved "once and for all"—something indeed strongly suggested, but not advocated, by the information-processing approach represented by Newell and Simon (1972). This leads many authors to consider that design is not problem solving, opposing "problem solving" to "constructive" and "creative" activities. Stolterman (1991), for example, opens the abstract of his essay "How System Designers Think About Design and Methods. Some Reflections Based on an Interview Study" with the following words: "Today system design methods seem to presuppose the system design process as problem solving, i.e., as repair of a malfunctioning reality." Later on in his paper, he completes this stance as follows: "The design process should instead be viewed as a creative way to design a new reality." In an interview with Hartfield, Kelley (1996) uses the term "problem-solving" in the sense of "fixing something that is deficient or defective," and concludes: "Design is not problem-solving" (p. 153). For Winograd (1996), design is not problem solving either. "Design is creative"; "[it] cannot even be comprehended as *problem solving*" (p. xxiii). To this respect, he opposes design to engineering: Engineers are problem solvers, not designers.

In the situativity community (see later discussion), the term "problem" is also the object of many comments. "Situated cognition research calls into question what constitutes a problem for a person. . . . Problems in cryptarithmetic and the Tower of Hanoi, as 'situations' given to a 'problem solver,' are contrasted with the manner in which a person experiences a problematic situation," writes Clancey (1993, p. 104).

A "Problem" in Cognitive Psychology: One's Representation of a Task

These approaches to "problem" and "problem solving," however, do not really address the classical cognitive-psychology view of these entities, which does not defend the positions criticized. Analyzed from a cognitive-psychology viewpoint, all that is difficult or deficient does not necessarily constitute a "problem." In cognitive psychology, "problem" is a technical term with a precise definition: It qualifies people's mental representation of their task.

We adopt Mayer's (1989) definitions based on his distinction between two types of problems, routine and nonroutine problems. A task is a "routine problem" *for someone*, if *the representation this person has constructed of the task*, although not eliciting a memorized answer, evokes a well-known procedure that, once applied, allows the problem to be solved. For "nonroutine problems," a person's task representation does not evoke a procedure, so that the person must construct one. For most adults, 863 x 725 will be a routine problem, whereas it will be a nonroutine problem for nearly every young child. The problem character of a task is thus a relative feature that depends on the task situation and on the person who must deal with it: A task that constitutes a problem for one person does not necessarily do so for another person (Leplat, 1981).

Several authors in the SIP tradition do not distinguish routine and nonroutine problems (Hoc, 1988a; Richard, 1990). They consider that if people's task representation evokes a procedure allowing them to attain their goal, the task constitutes no "problem" for them. Notice that some authors distinguish a "narrow view of problems" from what they consider a broader one, referring, for example, to "all situations where a need for action exists" (Ö. Akin, 1986a, p. 228, an author in the information-processing tradition).

In this book, we recurrently use the term "problem" where "task" would have been more appropriate in our opinion. This is the case in the presentation of theoretical approaches adopting this use (especially the SIP framework), but also in historical and more informal presentations where the term is commonly used by the authors we refer to.

Design tasks comprise both routine and nonroutine subproblems.[1] This book focuses on the nonroutine problems in design projects. From the cognitive viewpoint that we adopt, routine problems are not very interesting. From a sociocognitive viewpoint, they might be: Interaction is not restricted to a particular type of problem.

We wish to emphasize that an expression such as "design is not problem solving" is an abbreviated form of "many design tasks constitute no problem-solving tasks for the designers in charge of these tasks."

The Double Status of "Problem" and "Solution"

Design starts with requirements, generally provided by a client (even if one may of course retrace the source of these requirements to people's needs and wishes). In a rather imprecise and incomplete way, they specify the artifact to be designed. The *problem* representations constructed by designers on this basis are to result in a *solution*, that is, into specifications that specify precisely and in detail how to implement them into the artifact. Analyzed in terms of a problem-solving path that consists of a transformation from the initial problem representation into the representation of a solution, the path comprises a great number of intermediate steps leading to a great number of intermediate states, each with a double status. Until the final implementation level is attained, that is, as long as a solution is not yet detailed and concrete enough to specify the implementation of the artifact, each solution developed for a problem constitutes itself a new problem to be solved. That is why the same entity may be referred to as a "solution" or as a "problem," depending on whether it is an output or an input to a problem-solving action. If we want to insist on this relative character of problems and solutions, we use the term "problem solution" (Visser, 1991b).

The emphasis that we put on the representational aspects of problems and designing is generally absent in the classical cognitive design research literature. Nevertheless, the term "design problem" in this literature, first, refers to a designer's starting point for a design project, without any assumption concerning

- predefined, preexisting features that would discharge the designer from, for example, constructing a representation of the problem (the problem is not "given").
- the problem representation being constructed once and for all (i.e., at the beginning of the project: a designer does not proceed to the construction of *the* problem representation).

Second, the term "design problem" also refers to later "solution" states elaborated based on the initial requirements. It indeed refers to all representations of the design project on which a designer continues to work. As long as the designer does not consider the design project as "finished," the "design problem" is not yet "solved"!

---

[1] As "problem" and "subproblem" are relative notions (like "solution" and "subsolution"), we use the term "problem" (and "solution") unless there is a risk of confusion.

In engineering tasks, the client's requirements generally specify the artifact to be designed in terms of a goal to be attained under certain functional and often material constraints. An example is an automatic installation that is to manufacture certain automotive parts, at a certain speed, and with a certain result (Visser, 1990), or a satellite that is to fulfill a certain mission, with a certain useful charge (Visser, 1991b).

Even if the ultimate solution of the design process consists of the *specifications* of an artifact — thus, it is not the artifact product itself — this solution specifies a *concrete* solution, outlining *how* to *implement* the specifications. Before this ultimate solution is proposed, the global design process often proceeds by analyzing and further specifying the requirements, in order to transform the goal in terms of a *concept* or a *function* that allows attaining this goal.

On intermediate levels, the same progression may be observed, namely from a goal, via a concept or a function, to its implementation. At these levels, a problem, defined in terms of a goal to be attained, may be solved immediately in concrete terms — a situation never observed at the global design-process level. At the level of subproblems, there may be many routine problem-solving situations in design!

Problem "Difficulty": A Question of "Cognitive Cost"

In a context where problem solving is analyzed from a cognitive viewpoint, "difficulty" is not a synonym of "problem." A task can be more or less difficult, that is, can require more or less processing effort (Gilhooly, 1989, p. 3). From a cognitive-psychology viewpoint, this effort may correspond to what we have been calling elsewhere "cognitive cost." We have operationalized the cognitive cost of a design action in terms of "the number of information units to be processed and the nature of this processing (both the cost of accessing the required information, and that of using it)" (Visser, 1994a, p. 260).

Given the relative character of problem "difficulty," it is not evident to compare design activity in experimental tasks with the activity implemented in an industrial design project that takes weeks to complete. Many problems used in experimental studies are rather simple, and take between 1 and some 2 hours to complete. Their authors qualify them as, for example, "fairly straightforward by professional standards, however [demanding] a significant degree of problem-solving and design behaviour" (Davies, 1991a, p. 179), "novel," "challenging to" the "subjects"[2] (Adelson & Soloway, 1988), "familiar and practiced" (Byrne, 1977), "novel," "because none of [the] designers had designed an identical system in the past, although they had worked on related . . . systems" (Guindon, 1990, p. 314). There are, of course, exceptions: Subjects in Ullman et al.'s (1988) study take between 6 and 10 hours for a "fairly simple, yet realistic, industrial design problem."

---

[2] The term "subject" is no longer "politically correct": "Participant" has replaced it. According to the authors and the historical context, we use one or the other term.

Processes, Operations, and Activities.
Stages, Phases, and Steps

Before our presentation of the "process" of "problem solving" and its component "stages" or "steps," some nuances concerning these and related notions may be appropriate.

The term "process" is being used in various contexts in the literature, two of which are distinguished and used in this book. In the expression "design process" or without any further qualification, we use the term in the same context as in engineering and design-methodology communities, namely as referring to all action globally taking place during a design project.

Qualified by the attribute "cognitive," it concerns a smaller entity. The term "cognitive process" generally refers to a psychological basic mechanism,[3] such as memory activation, storage, encoding, matching, establishment of a relation between two basic units—but the term is also used for cognitive operations, which are more global (see later presentation).

"Activity" is generally used for a collection of cognitive processes spread out in time, but endowed with a unit and an organization (see Le Ny, 1991, in Bloch et al., 1991). The cognitive entities that correspond to these activities can be more or less fine-grained. Clear examples of rather coarse-grained entities—thus "activities"—are the three classically distinguished design activities: Construction of representations, solution generation, and solution evaluation. However, the distinction between activities and operations is not always obvious and we generally use the term "activity" for both, from such high-level cognitive activities to lower-level entities that might be qualified as "operations."

Indeed, between the global design process and the basic processes, one may distinguish "operations" at more or less fine-grained levels of activity. "Operation" also has a connotation of nonintentionality: One performs "operations" on mental images.

In AI literature, the term "method" is used. We do not assign a particular technical sense to this term, and use it exactly as their authors do.

"Step" is sometimes used as a nontechnical term when the nature of the underlying action is not relevant (basic process, operation, or activity).

When used in a technical sense, we adopt the term "stages" for prescribed steps, and "phases" for periods in the activity that can be observed to actually constitute entities that are somewhat separate from a cognitive viewpoint. This means that stages *can* correspond to phases (and vice versa)—but they will seldom do so.

---

[3] The cognitive-psychology literature also uses the terms "operator" or "operation," which we do not use in the context of a mechanism. In the SIP literature (especially Newell & Simon, 1972), "operator" is used for one of the problem components.

"Problem Solving":
A Global Process Versus a Particular Stage

As used until now in this section, the term "problem solving" refers to a global process, which actually takes place through a big number of small steps, "from problem detection through various attempts to problem solution or problem abandonment" (Gilhooly, 1989, p. 5). These different steps may be characterized as "problem forming" or "problem finding," to use two distinctions used by Simon (1987/1995c; see also Thomas, 1989), "problem attempting," "problem formulation" (Gilhooly, 1989; Thomas, 1989), "problem setting" (Schön, 1983), "problem conceptualizing," "problem framing" or "reframing" (Schön, 1988), and "problem defining" or "redefining" (see also the list of stages enumerated by Thomas, 1989, presented in the section on process models).

In addition to the problem-solving methods of heuristic programming (generate-and-test, hill climbing, match, heuristic search, and induction), Newell (1969) identifies, as "the many [other] parts of problem solving" (p. 407), recognition, evaluation, representation, information acquisition, method identification, method construction, executive construction, and representation construction.[4]

It may be noticed that authors adopting the information-processing approach to problem solving may themselves have contributed to the possible misunderstandings surrounding the notion of "problem solving" (see our remark that the view of a "problem" as a task that can be resolved "once and for all" is strongly suggested by the information-processing approach). At some occasions, these authors have presented what we qualify here as "problem-solving steps" as distinct from "problem solving." So did Simon (1987/1995c) in his lecture titled "Problem Forming, Problem Finding, and Problem Solving in Design." The general spirit, however, of the cognitive-psychology approach to problem solving is the following: Make part of "problem solving" all activities leading from a problem's specifications—however incomplete and imprecise they may be, as generally is the case in design project requirements—to its solution—be it temporary or definitive.

Newell and Simon's
Symbolic Information Processing (SIP) Approach

Only a very brief characterization of the SIP approach to problem solving in general can be presented in this book, but it is further discussed in the context of Simon's approach to design. From the late 1950s, early 1960s on, the SIP approach had been developed by Newell and Simon, together with colleagues such as Shaw (Newell, Shaw, & Simon, 1958; for other references, see Newell & Simon, 1972, p. 791, Note 1) (for a succinct presentation of the SIP approach to problem solving, see, e.g., Simon, 1978, 1979).

---

[4] Among these "methods" and "parts" are actions that we qualify as "strategies," others as "activities" or "operations," and still others as "elementary processes."

Newell and Simon's (1972) definition of a "problem" corresponds closely to Mayer's definition of nonroutine problems that we have adopted. Yet, the authors do not distinguish routine from nonroutine problems. Newell and Simon consider that "a person is confronted with a problem when he wants something and does not know immediately what series of actions he can perform to get it" (p. 72) (see also Hoc, 1988a; and Richard, 1990, who adopt this definition). Newell and Simon's position here is very close to that of the psychologist Karl Duncker, proponent of Gestalt psychology, which is a rather different approach to cognition than the one adopted by SIP. Duncker indeed considers that "a problem exists when a living organism has a goal but does not know how this goal is to be reached" (1945, quoted in Gilhooly, 1989, p. 2). Gilhooly offers an information-processing rephrasing of Duncker's definition: "A problem exists when an information-processing system has a goal condition that cannot be satisfied without a search process" (p. 2).

Newell and Simon analyze problems in terms of "states" and "operators." A problem consists of an "initial state," a "goal state," and of a set of "operators" that are "legal," in the sense that they may be used for transforming the initial state into the goal state. Operators have constraints that must be satisfied before they can be applied.

Problem solving takes place in a "problem space" that contains these states, operators, and constraints. Confronted with a task, a problem solver "represents the situation in terms of a *problem space*, which is his way of viewing the task environment" (Simon, 1978, p. 272). Search plays a central role in the SIP approach to problem solving. "The search for a solution is an odyssey through the problem space, from one knowledge state to another, until the current knowledge state includes the problem solution" (Simon, 1978, p. 276).

The notion of "problem space" also has led to many misunderstandings. In Newell and Simon's (1972) terms, "problem space" is a concept that covers problem solvers' representation of their task. It seems, however, easy to forget that a "problem" is not an entity that exists "out there," "given" to a person in order to be solved. Researchers often overlook the fact that a "problem" is a particular representation that an individual person constructs of a task whose presentation (verbal and/or written, implicit and/or explicit) is open to multiple interpretations. This leads to authors describing the "problem space" as defined by the problem, or as being the problem's decomposition into components—in short, as an objective entity that exists independently of a particular person who is confronted to a particular task.

A second misunderstanding is to confound the "problem space" with the "basic problem space" and to consider that the problem space contains, or corresponds to, all possible problem states. The basic problem space is a theoretical entity, which Simon (1978) defines as "the simplest problem space for a task . . . [which] consists of the set of nodes generated by all legal moves" (p. 276).

Several aspects of this approach are discussed later, especially the roles attributed to search and to representation.

4.2.   DEFINITIONS IN THE DOMAIN OF DATA COLLECTION

The appropriateness of a data-collection methods depends on the goal that one has in data collection (hypothesis testing, comparison, evaluation, or exploration) and the nature of the activity one is interested in (see Gilmore, 1990b).

Data Collection in the Psychological Laboratory Versus in Designers' Normal Work Situation

We are, of course, conscious of the tension and the difficulty of attaining equilibrium between two types of risks to diminishing the ecological validity of a model as they are related to two types of data-collection situations:

- Experimental situations that may not be representative of the real design situations that we are heading.
- Work situations (the "field") that may lead to results whose validity is restricted to these particular situations.

Data presented in this book as materials for a cognitive model of design come mostly from field studies of design activities, even if many characterizations of design that are presented in the literature do not have such an empirical basis (e.g., Simon's model of design).

As we claimed in Visser and Hoc (1990, p. 239), the methodological choice of observations on large, professional design projects is essential. More often than not, experiments are based on small design tasks, and use problem statements without any systematic selection. In such conditions, the ecological validity of the results is questionable. Nevertheless, the size of a design project is not the only relevant dimension to predict the problem character of the designers' task. A large project, if it constitutes a task that is familiar to a designer, can take a long time to design, without involving much problem solving. Conversely, a small-size project can accurately represent some problem-solving processes that are typically used in the development of large projects. As observations on small-size tasks are more easily conducted on several designers, such tasks are particularly appropriate to evaluate individual differences or to obtain generalizable results. When one reduces the size of a task, certain features of large, professional design projects are, however, difficult to reproduce (e.g., working-memory management, the role of the future artifact's user, or the role of different information sources).

In this context, the "cognitive paradox" (Dunbar, 1999) is particularly illustrative. Classical cognitive-science laboratory experiments repeatedly have demonstrated that people experience major difficulties with respect to many forms of reasoning (analogical reasoning, causal reasoning, hypothesis testing, or generalization). However, many of the biases identified in the laboratory are absent when people reason in their habitual environment. As regards analogical reasoning, that is, the example analyzed by Dunbar, the author argues that this cognitive paradox is not due to professionals having greater expertise than experimental subjects and thus using the relevant structural information that these subjects do not use. Dunbar shows that it is also possible for novices in the cognitive laboratory to use structural relations without any training. He attributes this result to features of both people's tasks and representations. The author indeed shows that important characteristics of both their task and their underlying representation of a concept reliably predict whether people will use structural or superficial features when making an analogy (Dunbar, 1999).

Explorational Studies

This approach is often used to answer "open" questions, like "How do people X?" or "Which issues are involved by doing Y?" The method may also be called "observation" (e.g., as the first moment in the "experimental cycle"— observation, hypothesis, verification—proposed in 1865 by Claude Bernard in his "Introduction à l'Etude de la Médecine Expérimentale" [Introduction to the Experimental Study of Medical Science]). In this book, the term "observation" is used for a particular type of explorational method, presented later.

"The prime difficulties with exploratory data collection are that it can be very difficult to collect and record such data and it is rare for the resultant data to be well-defined" (Gilmore, 1990b, p. 86).

As proposed by Claude Bernard, exploration generally is the first phase of research in a new domain—which design still constitutes, even after some 30 years of empirical studies. A consequence is that, because much cognitive research on design has been conducted up to now with the aim of exploring rather than of hypothesis testing, many studies have led to conclusions that are "only" hypotheses for further study—no unshakable "truths" or "facts."

We consider that observations, be they exploratory or not, are "guided" by a theoretical framework, even if it is only sketchy and provisional. Not all our colleagues share this position. Especially situativity-inspired researchers often adopt other viewpoints. Nevertheless, Bucciarelli (1988), famous for his inclination to situativity-related ideas, also considers that "what one sees as designing . . . depends upon one's interests and perspective" (p. 159).

Example. In our first studies of software designers (Morais & Visser, 1985; Visser, 1985, 1986a, 1986b, 1986c; Visser & Richard, 1984), we did not yet realize the importance of reuse. At the time, our theoretical framework did not yet include this notion, and so we did not "observe" reuse! We did not classify the solutions reused by the designers as such—even if the designers themselves did refer explicitly to these solutions as "old solutions," or as solutions "already used in the past" (or other, comparable idioms).

Two particular explorational methods are referred to in this book, namely observational studies combined with simultaneous verbalization, and interviews. With respect to the data-collection methods specifically appropriate to cognitive design studies, more examples and details can be found in Falzon and Visser (1989), Visser and Falzon (1988, 1989), and Visser and Morais (1991).

Observational Studies

In observational studies, one generally wants to collect data concerning as much aspects of the activity as possible. In cognitive design studies, one wants to collect data on both internal and external aspects, both production and communication aspects of design.

In controlled studies of small tasks, the whole design session can be recorded, via audio or video (Cross, 1997; Cross et al., 1996). In field studies, however, conducted on industrial design projects that take weeks, if not months, such exhaustive data collection is generally not realistic, though it is possible to do it: The data *can* be collected, but their analysis would be unfeasible. Therefore, in that case, researchers often take notes on designers' actions and collect all documents that the designers produce during their work.

One may, of course, gather data selectively in a field study that focuses on a particular component or aspect of design. This requires, however, that one already have a kind of model of the global activity so that one knows when the particular component or aspect is supposed to appear in the global process.

The following presentation of data collection is based on our functional-specification study, an observational study conducted on mechanical design (see Table 6.1). The same approach was adopted in our two other early design studies (software design and software debugging; see Table 6.1).

Notes may concern:

- The designers' problem-solving actions (disclosed by their verbalization), their other remarks and comments.
- The order in which designers produce the different documents, and how they gradually build them up.
- The changes they make.
- The information sources consulted, both colleagues and documents.
- The events considered by the observer to be indicators of the designers encountering difficulties.

Documents collected may be:

- The different versions of the plans, schemas, and other external design representations, both the intermediate[5] and the definitive versions, and both

---

[5] Each evening, when leaving, we generally collected all paper from the trash bins in designers' offices!

the versions for the client and those that designers construct for themselves during their work.

• Photocopies of all documents consulted.

Verbalization Versus Introspection

In addition to note taking and document collection, researchers often ask the people they observe to "verbalize their thoughts" or to "think aloud." This means that they ask them to report all mental activity with respect to the information they take into consideration, the choices they are confronted with, the criteria used to take a decision, their reasoning, their hesitations, their questioning past decisions, and so forth. Recording these "verbalizations" leads to a protocol whose analysis has been spelled out in great detail by Ericsson and Simon (1980, 1984/1993).

There are essential differences between "verbalization" and a certain form of "introspection," that is, the method frequently used around 1900, in the beginning phase of psychology—and still often adopted by people developing methods or assistance tools (working in the domain of design methodology, software engineering, or HCI).

This introspection consists in *commenting on* one's mental activity, thus, not verbalizing this activity, but talking *about* it. What is asked to be expressed in studies applying verbalization—which may be simultaneous (concurrent) or retrospective (Gero & Tang, 2001)—is not comments on one's activity, but the information one attends to and that has been generated by the task-directed processes underlying the activity—insofar as this information may be made explicit, which is not always possible. If researchers are interested in a mental activity— for example, the cognitive activity underlying designing—, they do not want to collect designers' comments, which may translate these designers' opinions about the mental activity in question, or rationalizations of this activity. The data that design researchers want to obtain is the *direct* traces of the information used in the mental activity, which are *indirect* traces of the internal cognitive processes underlying the activity. In order to be exploited in descriptions or models of the cognitive activity in question, this data is to be analyzed, according to strict rules and methods. This analysis constitutes a necessary, subsequent step that cannot be replaced by "subjects [speculating] and [theorizing] about their processes. . . . [One wants to leave] the theory-building part of the enterprise to the experimenter. There is no reason to suppose that the subjects themselves will or can be aware of the limitations of the data they are providing [when probed to proceed to introspection]" (Ericsson & Simon, 1980, p. 221).

Notice that, even among researchers using protocol analysis themselves, there seems sometimes to be confusion concerning the status of the data collected. Some of the misunderstandings encountered in the research literature concern the following three points:

- *Data collected: Direct or indirect traces of the activity.* Some researchers seem not to be aware that the data collected does not constitute a direct trace of the mental activity they are interested in, but of the information used in the activity, and that the nature of the activity is to be inferred on the basis of this information, its form, structure, order of expression, and so on.

- *The communicational character of verbalization.* Verbalization in this context does not constitute an interactional activity, in the sense of a dialogue or conversation. The researchers are supposed to provide the designers with such instructions that they do not consider themselves involved in a communication with the researchers who are observing them.

- *The risk of rationalizing the activity.* Interviews risk leading the interviewees to rationalize, in their description, the way in which they *think* to proceed. However, if designers are appropriately instructed, there is no reason why concurrent verbalization would lead to such rationalization of their activity: People instructed to verbalize do not talk *about* their activity.

In our studies examining planning and organization of the design activity, we have revealed the risks of introspection (Visser, 1990). We have shown indeed that the plans presented by designers as translating the organization of their activity did not coincide with the actual organization of their activity (see the example presented at the end of the section *Interviews*). It would therefore be questionable to use such descriptions based on introspection, as data representing the actual organization of the interviewees' activity (see also Malhotra et al., 1980).

The technique generally adopted by cognitive psychologists in order to analyze such verbalizations is "(verbal) protocol analysis" (Ericsson & Simon, 1980, 1984/1993). Ericsson and Simon's claim that, if the conditions are satisfied, protocol data can be collected during task execution without interfering with task performance, has been criticized, often by reference to a paper by Nisbett and Wilson (1977). In our view, Ericsson and Simon, in the revised 1993 edition of their reference work published in 1984, have countered Nisbett and Wilson's objections.

The use of verbalization is submitted to several conditions. Ericsson and Simon (1980) summarize the effects that the instruction to verbalize may have on the cognitive process that the researcher is interested in. This summary expresses the two most important conditions.

"Producing verbal reports of information directly available in propositional [that is, language or verbal] form does not change the course and structure of the cognitive processes. However, instructions that require subjects to recode information in order to report it may affect these processes" (Ericsson & Simon, 1980, p. 235).

"Only information in focal attention can be verbalized" (Ericsson & Simon, 1980, p. 235). "Automation . . . [makes] the intermediate products [of task-directed information processing] unavailable to STM [short-term memory], hence unavailable for verbal reports" (Ericsson & Simon, 1980, p. 225).

Thus, according to the nature of the target activity, verbalization is a more or less appropriate method. It would not be useful, for example, to obtain data concerning the knowledge used by Dutch people when cycling, because this knowledge is automated.

Protocol analysis is a particularly time-intensive activity for a researcher. There have been estimates according to which it may take some 10 hours of protocol analysis for every 1 hour of collected protocol (Eastman, 1999). Referring to the normal magnitude of common design projects, Eastman judges that "it is clear that we will never collect one large scale design protocol" and that other types of approaches are needed. He considers that sampling strategies are to be applied on design projects.

Verbalization and analysis of the resulting protocol are often used in "clinical" studies, that is, in analysis of the activity performed by only a few, or even only one single subject, with the ensuing risks concerning the possibility to generalize the results. As claimed, however, by Anzai and Simon (1979) with respect to this type of work, "[even though] one swallow does not make a summer, . . . one swallow does prove the existence of swallows. And careful dissection of even one swallow may provide a great deal of reliable information about swallow anatomy" (p. 136).

Since the early days of research on design activities (e.g., Eastman, 1970), many, if not most, empirical design studies, especially the cognitive ones, use simultaneous verbalization (Ö. Akin, 1979/1984; Baykan, 1996; Eckersley, 1988; Galle, 1996; Gero & McNeill, 1998; Letovsky, 1986; McNeill, Gero, & Warren, 1998; Suwa & Tversky, 1997).

The request to "verbalize one's thoughts" or to "think aloud" is necessary only for data collection on individually conducted activities. People working together "naturally" express their thoughts. The analysis of the corresponding protocols requires, however, also a specific method.

With our colleagues of the EIFFEL research group (see e.g., Détienne & Falzon, 2001), we have developed the COMET method for the analysis of collaborative design processes (Darses et al., 2001), integrating protocol analysis as developed for the analysis of individually conducted activities, and pragmatic linguistics' verbal-interaction analysis (Kerbrat-Orecchioni, 1990, 1992, 1994). We did not introduce, however, the means to analyze linguistic phenomena such as modalization procedures, for example, expressions of addressing or of politeness (Araújo Carreira, 1997, 2005).

A new endeavor in which we have engaged recently consists in extending the analysis of verbal interactional data to that of other semiotic systems (nowadays, especially in an HCI context, often qualified as other "modalities"). At present, we have developed a description language for the graphico-gestural activities, and are examining the articulation between graphico-gestural and verbal dimensions in collaborative design interaction (Détienne, Visser, & Tabary, in press; Détienne & Visser, 2006; Visser, in press; Visser & Détienne, 2005).

Task Versus Activity

In a cognitive-psychology context, our focus on the *activity* of design refers to several opposing notions, especially that between "task" and "activity."

The concept "task" refers to either what people are supposed to do (i.e., their "prescribed" task, as it has been specified by their manager, by instructions or manuals), or what they set themselves as a task and what they actually realize (their "actual" task). These two coincide only seldom (Leplat, 1981).

"Cognitive activity" refers to the way people actually realize their task on a cognitive level: How they make use of their information sources—their knowledge, other information sources they refer to, and still other tools—in elaborating their intermediate and final productions.

In ergonomic analysis of work situations, authors generally distinguish three analysis levels:

- Task constraints or conditions that are external to people's activity.
- "Subject characteristics" or conditions that are internal to people's activity (e.g., their expertise).
- The activity itself, which translates the way in which the subject characteristics are implemented in order to respond to the task constraints (Leplat & Cuny, quoted in Leplat, 1986/1992a, p. 27).

Problem characteristics are often presented as an example of task characteristics. As defended previously, we consider that the characteristics of a "problem" first depend on the "subject" confronted with the task: Even "problem structure," often presented as a typical "problem characteristic," depends, in our view, on the knowledge that a person has in the corresponding task domain!

Interviews

The remarks formulated in the section *Verbalization Versus Introspection* with respect to the data that verbalization or introspection may, or may not, provide on mental activities, apply, *mutatis mutandis* to the data that one may obtain with interview-like methods.

Notice that "verbalization" is sometimes used as a general term, including every oral expression in experimental conditions, both answers in interviews and reports of mental activity as referred to previously. We restrict the use of the term to this last application.

The difference between task and activity can be illustrated as follows: Whereas interviews may provide data on somebody's task, they are generally not of much use with respect to many aspects that are essential to understanding one's activity. Interviews are generally the type of data-collection method used to develop interfaces or other interactive-system components. This therefore may lead to these systems being based on data that applies to people's task rather than their activity—and thus possibly not adapted to people during their activity.

Several types of interviews are possible, from the completely "open," undirected interview, which is nearly like a sociable conversation, to the completely structured interview, which is organized around a specific set of questions. The completely undirected interview is rarely used in a research context, because, even if researchers do not know which *particular* data they are going to obtain, they generally know the *type* of data to be collected, so they will structure their interview around certain types of questions.

Semidirected interviews are generally the first approach used in the study of a new domain of research. They allow data to be obtained on important aspects of a task such as its global description, its major stages, or phases, and the main information sources used. With respect to the activity, it risks inducing the interviewees to present it in a much more structured way than the actual activity to which (they think) they refer. The following example underlines once again the importance of the distinction between "task" and "activity."

Example. At the start of our longitudinal study concerning three stages in the design of IPC software (see Table 6.1), we interviewed designers about their activity; afterward we observed their actual activity. On the basis of the data concerning this actual activity, we concluded that the structure of the representation that the designers presented in the interviews as representing their activity did in fact not coincide with the structure of the actual activity (Visser, 1991a). The designers described their activity as being organized in a hierarchical way, but their actual activity was opportunistically organized—a characteristic of design activities that, since these early days, has been observed more and more in empirical design studies (Visser & Hoc, 1990).

Recently, Vermersch and colleagues have developed data-collection methods that might be qualified as situated "between introspection and verbalization" (Vermersch in Depraz, Varela, & Vermersch, 2003; Vermersch, 1994) and that allow collection of data on activity-related aspects.

5 Models of Design

In 1984, that is, some 20 years ago, Cross edited a series of papers presenting a history of the *Developments in Design Methodology*, collecting texts written by the authors who had been involved in the underlying "movement" (Cross, 1984a). The collection covered the 20 preceding years, that is, the period from 1962 (when the first Conference on Design Methods marked the birth of design methodology) until 1982 (when the Design Policy Conference signified the coming of age of design methodology). The organization of Cross' book reflected the progression of the movement through four stages:

- Prescription of an ideal design process (1962–1967: The period of "systematic design" proposed by the proponents of the movement concerned with "design methods"—in 1972 qualified by Rittel, 1972/1984, as the "first generation design methods").
- Description of the intrinsic nature of design activity (1966–1973: When design problems were discovered to be not so amenable to systematization. Authors tried to understand their apparent complexity, attributing it in large part to the "ill-structuredness" of design problems).
- Observation of the reality of design activity (late 1970s: Methodical collection of data on the actual design activity).
- Reflection on the fundamental concepts of design (1972–1982: Emergence of a more fundamental and philosophical approach to design method).

Given these developments shown by Cross, one might expect that, today, some 40 years after several of Cross' historical design studies, we are in a new stage, beyond "simple" descriptions and observations. One might expect that predictive models have been built and compared, and that a consensus has been attained concerning "the nature" of design. This is, however, not the case at all!

Atwood, Gross, McCain, and Williams (2004) in "Science of Design: Why We Need It and Why It Is So Difficult to Achieve" advance the notion that initiatives undertaken until now by people to build a theory of design have been doomed to failure." We, as a profession that holds design skill in high esteem, cannot achieve a theory of design because we, as designers, do not have a common view of what design is" (Atwood et al., 2004, p. 1). The authors propose an approach that is not based on classifications that use primarily "the analysis of one or a few people." They aim to identify the classification established by "the global community of designers," in order to find out "how the various views of design are related and to use this as a basis for building an overarching theory of design" (Atwood, McCain, & Williams, 2002, p. 126).

Through a bibliographic cocitation analysis (for a technical explanation and references, see the authors' paper), Atwood et al. (2002) identify what they consider the way in which "the design community thinks about design" (see the title of the paper). The results of a hierarchical cluster analysis are a classification into seven "key author clusters" representing seven "identifiable theory groups or schools of thought/practice in design" (p. 125). The clusters gather around "design complexity," design taxonomists," "design rationale," "participatory design," "user-centered design," "design theorists," and "cognitive engineering." This result confirms the lack of a "common view": "There is no central focus that holds the design community together" (p. 130).

Two other interesting aspects of the clustering underlined by Atwood et al. (2002) are that "some views of design focus strongly on people, others do not," and "some views of design build theory, some want to build successful systems" (p. 131).

This chapter presents the major types of design models developed and adopted in practice and in research, classifying them into prescriptive and descriptive models, and stage and process models.

5.1.   PRESCRIPTIVE AND DESCRIPTIVE MODELS

Most cognitively oriented authors focus on descriptive models, which they oppose to prescriptive (or normative) models. Predictive models are not even yet envisioned in this domain.

Prescriptive Models

Prescriptive models of design are developed in order to monitor the design process. Generally linear and sequential, they stipulate that problem solving follows an abstract-concrete axis (generally corresponding to the implementation hierarchy, going from conceptual to physical specifications) by an iteration of two complementary stages. These are the generation of a solution (partial and intermediate), followed by its evaluation, leading to the generation of a better solution, which itself is evaluated again, and so on until the ultimate solution is obtained (see Darses, 1997, for a critique of this view; for references to older prescriptive models, see Hamel, 1995).

Prescriptive models of design and corresponding methods for engineering design are exemplified by Pahl and Beitz (1977, 1984).

Corresponding to prescriptive models and methods are norms. They differ according to countries and their cultures, working traditions, and otherwise: norm BS 7000 in Great Britain,[6] norm AFNOR X50-127 in France, norm DIN or VDI[7] 2221/2 in Germany (VDI, August 1987).

---

[6] "The Management of Product Design," published by the British Standards Institution in 1989 (according to Pahl & Beitz, 1996, p. xx).
[7] VDI = Verein Deutscher Ingenieure, that is, Organization of German Engineers.

Prescriptive models for software design. As in other domains of design, many models of software design have been proposed without any reference to the actual activity and are nevertheless used—or rather imposed—as the basis for managing software development. So are the waterfall model (Boehm, 1976), stepwise refinement models (Wirth, 1971), incremental release models (Basili & Turner, 1975; see also the IEEE Standard 1074-1997 for Developing Software Life Cycle Processes; and the ISO/IEC International Standard 12207-1995: Information Technology. Software Life Cycle Processes, Scacchi, 2001).

Methods proposed for software design are mainly of two types: Mathematical methods and structured methods. These different types of methods have been developed and are used mainly by two different communities. Because of their complementarities on several dimensions, integration of both is considered particularly useful by some authors, but has not yet been undertaken (Kitchenham & Carn, 1990).

Mathematical methods provide a notation that is supposed to be unambiguous, but these methods offer no rules or heuristics in software development, for example, for requirements formulation. They can be used for verification of software, but of course, only once this software is in an advanced phase of development.

Structured software design methods are manifold, as in other design domains (see, e.g., the ASE paradigm proposed by J. C. Jones, 1963/1984). Examples are data-flow analysis and design (Yourden & Constantine, 1978), data-structure methods (especially, Jackson Structured Programming, JSP, applied in Jackson System Development, JSD—see Jackson, 1975, 1983; but also SSADM, Structured Systems Analysis and Design Method, a standard for British database projects), Unified Process, and, particularly in the domain of database design, entity-relationship and object-oriented modeling. Contrary to the mathematical ones, these methods can be used for the specification of system structure.

The waterfall model as a continuing reference for software design. The Standard Waterfall Model for Systems Development, often abbreviated to the "waterfall model," is probably *the* classic software life-cycle model (Boehm, 1976, 1981, as quoted in many papers on design, e.g., Kitchenham & Carn, 1990; see also *The Standard Waterfall Model*, n.d.). This model defines a certain number of stages that the software development process is supposed to go through. According to this model, design is supposed to take place before coding and after requirements have been defined.

Descriptive Models

Few descriptive models of actual design are global models that cover the entire design process. Proposals generally focus on particular aspects (stages, strategies, activities) of the process. Darses (1994) has formulated elements for a model of design based on constraint management, focusing on solution development. Bonnardel (1992) has analyzed the evaluation component of design, which also relies on constraints or criteria. Burkhardt and Détienne (Burkhardt, 1997; Burkhardt & Détienne, 1995; Détienne & Burkhardt, 2001), and we (Visser, 1993c, 1999; Visser & Trousse, 1993) have contributed to possible models of reuse in design. We have also proposed elements for the organizational component of a model of design (Visser, 1988c, 1990, 1994a, where "organization" refers to the way in which designers themselves organize their activity).

In order to formulate global models of design that articulate its different components, one needs data about the different activities carried out by designers, their strategies, *and the articulation* of all these elements. Research conducted in cognitive psychology and cognitive ergonomics contributes to provide such data, generally through empirical observational studies. Such an approach to design is central to this book.

5.2.   STAGE MODELS
AND PROCESS MODELS

Most prescriptive models present design as a series of steps to be followed in a top-down, stepwise manner (Blessing, 1994; C. Warren & Whitefield, 1987). Indeed, prescriptive models generally are stage models—and most stage models are prescriptive.

There are, however, also descriptive, cognitive models that present design as proceeding in a series of consecutive steps. According to Simon's (1969/1999b, 1973/1984) model of design, ill-structured problem solving proceeds through two consecutive steps, first structuring the ill-structured problem, and then solving the resulting well-structured problems. This view is discussed in more detail later.

Another example is the descriptive model of architectural design developed by the psychologist Hamel (1995). The model that is task oriented rather than activity oriented and formulated at a macrolevel consists of three domain-specific components, namely analysis, synthesis, and molding—a triad that we encounter as the ASE stage model, with evaluation in place of molding.

It is not always easy to distinguish such descriptive models from cognitive-psychology models that present a global decomposition of the activity into "stages" with the *analytical* aim of globally rendering the activity. An example is the analysis of problem solving in "construction of a problem representation," "solution generation," and "solution evaluation" (see, e.g., Reimann & Chi, 1989; Richard, 1990; Visser, 1991b). Nevertheless, such decompositions that are useful on an analytical level generally do not render the actual activity, which is rarely structured into separate stages, be it two, three or four.

A comparable series of three stages is the cycle perception (or observation), trying, and evaluation, adopted, for example, by De Groot (1969, quoted by Roozenburg & Dorst, 1999, p. 37). Roozenburg and Dorst (p. 37) relate this cycle to the one proposed by Schön, namely naming, framing, moving, and evaluating. They compare it also to the stages adopted in "many different models of practical problem solving, and notably in the problem-solving model of System Engineering, which has become part and parcel of traditional systematic approaches to designing" (p. 37). The authors refer to "problem-solving cycles like the one in the VDI Guideline 2221 which comprises problem analysis, problem definition, system synthesis, system analysis, evaluation, and decision making" (p. 37, Note 7).

We present these different proposals in Table 5.1.

Notice that authors presenting models based on stages do not always make explicit whether they consider their proposals as representing the *actual* design activity. Roozenburg and Dorst (1999), for example, establishing a relationship between Schön's model and system-engineering models, imperceptibly glide from descriptive to prescriptive models.

Stages 2 and 3 can be related to the generate and test stages in generate-and-test approaches to design.

Stage Models

These models present the design process as a series of steps that are traversed consecutively, in a particular order. When translated into a method (a prescription), each stage must be finished before the next one is engaged, and once a stage is finished, one cannot come back to it.

Design stages have been defined as "[subdivisions] of the design process based on the state of the product under development" (Blessing, 1994, p. 10), so they are generally characterized in terms of their input and output. Stage models

TABLE 5.1

Confronting the Stages in Problem Solving Proposed by Different Models

Author or Source	Stage 1		Stage 2	Stage 3		
Simon	structuring (if ill-structured problem)		problem solving			
Hamel	analysis		synthesis	molding		
classical cognitive-psychology models	problem-representation construction		solution genera-tion	solution evaluation		
Schön	naming	framing	moving	evaluating		
De Groot	perception/observation		trying	evaluation		
Jones	analysis		synthesis	evaluation		
VDI Guideline 2221	problem analysis	problem definition	system synthesis	system analy-sis	evalu-ation	decision-making

usually do not describe the activity implemented in each stage, and that leads from its input to its output. In each stage, varieties of "techniques" or "methods" appropriate to the main task are proposed. Stages may be decomposed into finer stages, which may correspond to what *may seem* activities (classified as "operations" by Blessing, e.g., pp. 10, 20). The finer stages may be articulated in an iterative manner, that is, in cycles.

Some stage models do not merely cover the design process, but embrace the artifact's global life cycle, from the "policy stage," passing through the macrostage of "design," to the use or disposal of the product (for a list of stage models and their characteristics, see Blessing, 1994, p. 20).

In the engineering design domain, many models in terms of stages have been proposed. A famous example and reference in the mechanical engineering design domain is the prescriptive four-stage model by Pahl and Beitz (1977, 1984) (this model is similar to the one proposed by French, 1971):

- Product planning and clarifying the task (elsewhere often called "analysis").
- Conceptual design.
- Embodiment design.
- Detail design.

The third phase has received many other appellations: "Layout design," "main design," "scheme design," "draft design." French (1971) calls it "embodiment of schemes." The output of this third phase is a technical description, often in the form of a scale drawing, which, depending on the particular company involved, may be called a "general arrangement," a "layout," a "scheme," a "draft," or a "configuration drawing."

Other authors add a fifth and a sixth stage:

- Evaluation and decision taking.
- Presentation of results.

In their 1988 norm X50-127 "Recommandations pour Obtenir et Assurer la Qualité en Conception" [Recommendations for Obtaining and Assuring Quality in Design], the French norm-editing organism AFNOR distinguishes three stages: Feasibility study, preproject, and project development (see Perrin, 1999, p. 22).

Another famous reference in the engineering design literature is the ASE paradigm, which distinguishes three stages, namely analysis, synthesis, and evaluation. It constitutes the basis of the "systematic design" method, which is widely accepted in the design methodology community.

The reference for the analysis-synthesis distinction in design is J. C. Jones (1963/1984) (see also Asimov, 1962). Jones was one of the first authors to propose a way of organizing the design process so that logical analysis and creative thought—assumed by Jones to be both necessary in design—would each proceed in their own different way. By providing systematic methods for keeping data, information, requirements, and so on, outside one's memory, Jones' systematic-design method attempts to leave the designer's mind as free as possible for random, creative ideas or insights.

Stage models are often qualified as "linear" models (as opposed to "nonlinear" models). In two-stage linear models, the design process generally is divided into problem definition and problem solution (or problem structuring and problem solving). Buchanan (1990) qualifies them as follows:

> Problem definition is an *analytic* sequence in which the designer determines all the elements of the problem and specifies all the requirements that a successful design solution must have. Problem solution is a *synthetic* sequence in which the various requirements are combined and balanced against each other, yielding a final plan to be carried into production. (p. 355)

Buchanan (1990) presents the wicked-problem theory developed by Horst Rittel in the mid-1960s as an example of a nonlinear model, an alternative to the linear, step-by-step design models of that era (see Cross, 1984a).

Process Models

Process models present activities or operations supposed to be carried out by a designer in order to realize a task. Most cognitive models can be qualified as "process models," but their authors generally do not use this terminology. Process models are often the work of researchers from engineering design domains.

As a result of her comparative analysis of the principal prescriptive models, the industrial design engineer Blessing (1994) proposes a "process model" as the most appropriate basis for an AI-inspired support system. For each element of the "product" (each subsolution in a problem-solving framework), it combines stages and "activities." In a proposal that may be analyzed as a combination of Newell and Simon's (1972) two stages and Pahl and Beitz's (1977, 1984) four stages, Blessing distinguishes three main stages:

- Problem definition, resulting in a problem definition and a set of requirements.
- Conceptual design, resulting in a concept (or solution principle).
- Detail design, resulting in a full product description.

"Often [the detail design stage] is divided into embodiment design, in which the concept is developed resulting in the final layout, and detail design in which every component is fixed in shape and form, resulting in a full product description" (p. 10).

Notice that the model proposed by the psychologist Thomas (1989) seems in between the stage and process models. Thomas presents a long list of "stages" that problem solving goes through (see Table 5.2). He notices that his model is more a normative than a descriptive one. "In particular," he states that "problem

TABLE 5.2
Problem-Solving Stages Proposed by Thomas (1989) and Hypothetical Corresponding
Stages in Other Problem-Solving Models

Stages proposed by Thomas (1989)	Corresponding stages in other problem-solving models
problem finding	problem structuring
problem formulation - goal setting - goal elaboration	
idea generation	problem solving
idea evaluation - solution match with goal - solution match with environment	
idea integration	
solution acceptance or solution modification	
implementation - planning for implementation	(postdesign activities)
retrospective evaluation - measuring the outcome - evaluating the process	

finding, problem formulation, and looking back later at the problem-solving process are extremely useful but seldom utilized" (p. 325).

We suppose that problem finding and problem formulation in Thomas' model correspond to what in other models is called "problem structuring"; the stages from idea generation to acceptance or modification correspond to "problem solving"; and implementation and retrospective evaluation are activities that follow design, and are generally considered to be no part of design itself.

5.3.   CONFRONTING THESE DIFFERENT MODELS

Data from different sources concur in testifying that prescriptive models no longer have the power that for a long time has been attributed to them, or to their corresponding methods. Even if, in industry, they still possess a certain authority, both in the form of methods and as representations of design considered "factual," certain events put into perspective this influence. Such an occasion was the first Human Behaviour in Design symposium (Lindemann, 2003). In

2003, this symposium, organized in Pahl and Beitz's Germany with its particularly strong methodological tradition (based on prescriptive stage models), was attended not only by cognitively oriented researchers, but also by design methodologists.

Two questions can be asked in this context:

- Are the methods based on prescriptive models effectively followed, or not?
- If they are followed, do they lead to "better" design?

With respect to the first question, there are different types of evidence. Surveys of companies all over the world (Europe, USA, Japan, Korea) have shown that few enterprises have managed to control and improve their design process at the same degree of efficiency as they did for their production process (for references, see Culverhouse, 1995). Contrary to other phases of product development (especially production), there are no norms or company practices for controlling the duration of the design process.

On the basis of her analysis of prescriptive models, Blessing (1994) asserts that the corresponding normative methods are seldom really applied. She concludes that these models provide characteristics of ideal, rather than actual design processes.

Many other, cognitive studies of designers' activities confirm that the sequential organization of design such as prescribed by various methodologies, runs up against the actual activity of these designers. The cognitive processes that designers implement in their tasks do not follow such a sequence. The formal separation introduced in prescriptive stage models between problem analysis (upstream), and decision making and action (downstream) is in contradiction with the opportunistic character of the activity. This opportunism of design activity expresses itself, for example, in the intermingling of actual solution-generation and solution-evaluation activities (Visser, 1994a). The rest of this book abundantly presents results substantiating this claim.

If one wanted to analyze design as taking place in "phases"—that is, steps that differ with respect to cognitive characteristics—these phases would be very numerous.

With respect to the second question, there is little empirical evidence on both the quality of the design process and that of the resulting artifact (but see the later section *Quality of Design Solutions*). In a series of empirical studies conducted in both the field and the research laboratory, a group of German researchers has shown that designers using standard design methods did not produce better designs than their colleagues with long professional experience who did not use these methods (Pahl, Badke-Schaub et al., 1999).

Observance of the ASE Paradigm

Several types of criticisms have been formulated as to the ASE paradigm. "The processes involved in this view of designing use a terminology which is no longer widely accepted. The term 'analysis' has been replaced by 'formulation' or similar terms and 'synthesis' is now used to refer to a precursor of evaluation" (Gero, 1998c, p. 49). Even in the early days of design research, the fundamental critique had been formulated that the analysis-synthesis-evaluation cycle is inappropriate. Designers, rather then traversing three such stages, constantly generate new task goals and redefine task constraints. Ö. Akin (1979/1984) notices that "'analysis' is a part of virtually all phases of design. Similarly, 'synthesis' or solution development occurs as early as in the first page of the protocol" (p. 205). Designers generally start with a broad, top-down approach to the task. However, "all solutions do not arise from an analysis of all relevant aspects of the problem. Often a few cues in the environment are sufficient to evoke a precompiled solution in the mind of the designer. Actually, this is more the norm than a rational process of assembly of parts, as suggested by the term 'synthesis'" (p. 206). For other early critical reviews of prescriptive models, see Carroll and Rosson (1985), Cross (1984b), Dasgupta (1989, pp. 31–33), and Visser (1987a).

Observance of the Waterfall Model

As holds for stage models in other domains of design, the waterfall model does not reflect the actual activity of people involved in software design. This is not only a conclusion formulated by psychologists on the basis of empirical studies (Détienne, 2002a; Gilmore et al., 1994; Hoc et al., 1990; Visser, 1992b), software engineers themselves also formulate this judgment (Kitchenham & Carn, 1990; Löwgren, 1995).

In addition to the cognitively oriented remarks formulated in this chapter, some criticisms of the waterfall model that have no particularly cognitive nature are the following (*The Standard Waterfall Model*, n.d.):

- The model can be very expensive.
- Requirements are supposed to be fixed before the system is designed, but requirements generally evolve (cf. Carroll et al., 1997, 1998).
- Designing and coding often turn up inconsistencies in requirements, missing system components, and unexpected development needs.
- Many complications and technical hitches are not discovered until system testing.
- System performance cannot be tested until the system is almost coded, and hitches such as undercapacity may then be difficult to correct.

The waterfall model is also associated with the failure or cancellation of a number of large systems. As a result, the software development community has experimented with a number of alternative approaches, including Spiral Design (also proposed by Boehm, 1986, quoted in *The Standard Waterfall Model*, n.d.), Modified Waterfalls, (Evolutionary) Prototyping, and Staged Delivery (see McConnell, 1996, quoted in *The Standard Waterfall Model*, n.d.). In the last few years, a paradigm known as Extreme Programming has emerged that emphasizes reducing the cost of software changes, developing test cases before coding, developing code using pairs of programmers, and putting most of the documentation into the code (see Beck, 2000, quoted in *The Standard Waterfall Model*, n.d.). All these proposals have, however, not yet been evaluated from a cognitive viewpoint—even if they are becoming the object of empirical software-engineering studies (Jedlitschka & Ciolkowski, 2004).

One frequently comes across authors stating that the waterfall model has been replaced by an iterative approach to software development. At least two remarks may be formulated with regard to this:

- Such replacement may have occurred in the conclusions of research papers or in descriptive contexts, but the waterfall model still has a powerful position as prescriptive reference underlying design methodologies and software environments used in industry. It continues also to be "widely used on large government systems, particularly by the Department of Defense" (as noticed in *The Standard Waterfall Model*, n.d.).

- Presenting software development as an iterative process can only suffice as a first global characterization. Indeed, in order to characterize the process, one needs to identify, at least, the components of the cycles iteratively traversed, which often will be rather low-level, but also the conditions under which iterations occur (see, e.g., Malhotra et al., 1980).

6 Our Empirical Design Studies at a Glance

As we regularly refer to our studies for the presentation of examples, Table 6.1 presents our five main empirical studies of design at a glance. An overview of our early research on software design and mechanical design has been presented elsewhere (Visser, 1991a).

We refer to, at least, two other collaborative studies conducted with colleagues, not presented in Table 6.1.

In an empirical study, we observed the activities of team members during design-evaluation meetings (DEMs, also qualified as Technical-Review Meetings, TRMs) in a software development project that was to develop a business process simulator based on Petri Nets. The project required four full-time software engineers and lasted 1 year. The case study was based on the first 19 weeks of the project, the period needed to build the beta version of the product. An object-oriented paradigm was used throughout the development process. The meetings were required to be held before each document composing a milestone was accepted. The results presented in our publications (D'Astous, Détienne, Robillard, & Visser, 1998; D'Astous et al., 2001; D'Astous et al., 2004) are based on the observation and analysis of 7 representative meetings out of the 15 that were recorded.

Recently, we have conducted together with colleagues, in the context of the MOSAIC project (Détienne & Traverso, 2003), an observational study on an architectural design meeting (Détienne et al., in press; Détienne & Visser, 2006; Traverso & Visser, 2003; Visser, in press; Visser & Détienne, 2005).

TABLE 6.1
Our Main Empirical Design Studies at a Glance

Reference used in this book	Longitudinal industrial programmable controller (IPC) software design study			Composite-structure aerospace design study	Carry-ing/fastening device design ("Delft study")
	Functional specification	Software design	Software debugging		
Main publication	(Visser, 1990, 1994a)	(Visser, 1987b)	(Visser, 1988a)	(Visser, 1991b)	(Visser, 1995b)
First presented in (if different)	(Visser, 1986a)		(Visser, 1986b)	(Visser & Bonnardel, 1989)	(Visser, 1994c)
Also presented in	(Visser, 1987a, 1988c, 1989, 1991a, 1992b)	(Visser, 1986b, 1987a, 1988b, 1991a, 1992b)	(Visser, 1987a)	(Visser, 1991a, 1992b, 1992c, 1992d, 1993a, 1996)	
Task domain	Mechanical design: Functional specification of an automatic machine tool (AMT)	Software design: Control program for an automatic machine tool (AMT)	Software design: Control program for an automatic machine tool (AMT)	Composite-structure aerospace design: Unfurling antenna for a satellite	Industrial design: Device for carrying/fastening a backpack on a mountain bike
Routine-Nonroutine	rather routine	rather routine	rather routine	rather nonroutine	rather nonroutine
Designer	mechanical-design engineer	software engineer	software engineer	composite-structure designer	mechanical-design engineer
Working alone / in a team	foreman of a team	working alone	working alone	foreman of a team	working alone
Data collection	field study	field study	field study	field study	experimental study
Data collection	Observation and simultaneous verbalization: Note taking, document collection (selective)	Observation and simultaneous verbalization: Note taking, document collection (selective)	Observation and simultaneous verbalization: Note taking, document collection (selective)	Observation and simultaneous verbalization: Note taking, document collection (selective)	Observation and simultaneous verbalization: Videotaping and document collection ("exhaustive")
	Observations on engineer throughout his task of defining functional specifications for AMT's control part (IPC)	Observations on engineer throughout his task of developing the SW for an IPC using a declarative Boolean language	Observations on engineer throughout his task of debugging AMT's computerized control part (IPC)	Selective observations on designer during his task of designing an unfurling antenna	Observations on engineer throughout his task of designing a mountain bike attachment
Observation period	3 weeks (full time)	4 weeks (full time)	4 weeks (full time)	9 weeks (3–4 days/week)	2 hours

II

The Classical View on Design. The Symbolic Information Processing (SIP) Approach

In this part, we discuss the main classical view on design, which has been developed by Simon in terms of the more general symbolic information processing (SIP) approach.

Outline of this part. In the first chapters of Part II, we briefly present Simon and then characterize his approach to design. Afterward, we discuss this approach. We formulate nuances and critiques with respect to Simon's position. We also present Simon's "more nuanced" positions that he expressed in later texts, and we propose a possible explanation for the "simplifying" character of Simon's approach to design.

7 Herbert A. Simon

One of the first and main authors who have contributed to establish the foundation of a design theory from a cognitive viewpoint is Herbert A. Simon—who has analyzed design also from several other viewpoints. It may be, however, that this fame in design circles no longer holds among present-day students. Coyne (2005) notes that "the students in our architecture school are more familiar with Heidegger and Derrida than with Simon" (p. 13).

For many researchers from other domains, Simon is not especially renowned as an author in the domain of design, but as the winner of the 1978 Nobel Prize in economics. In AI, he is also a famous author. For many psychologists, Simon is, together with Newell, the founder of the main cognitivistic approach to problem solving: The symbolic information processing (SIP) approach (Newell & Simon, 1972).

It was also together that these two researchers received, in 1975, the "Nobel prize" in computer science, that is, the Turing Award (Newell & Simon, 1976). Newell and Simon met in 1952, at the Rand Corporation: "It was love at first sight at the intellectual level" according to Pitrat (2002a, p. 14). "Quickly they realised that they shared a common view on the functioning of the brain, symbol manipulator for Simon and information processing for Newell" (p. 14). Their first joint papers were published in 1956 (Newell & Simon, 1956; Simon & Newell, 1956). As a matter of fact, Simon's "interest in AI has been, from the beginning, primarily an interest in its application to psychology," as he declared in his 1995 IJCAI Award for Research Excellence lecture (1995a, p. 939).

For Simon, 1955 and 1956 were the most important years of his scientific life. At the end of 1955, his attention moved from administration and economy to psychology and computer science. This shift was due to the discovery that he made together with Newell and Cliff Shaw that the computer could be used as a machine to manipulate symbols (Pitrat, 2002a, p. 14) (see Newell, Shaw, & Simon, 1957a, 1957b). In 1958, the three together published their first cognitive-psychology paper in the *Psychological Review* (Newell et al., 1958). The year 1956 is also often considered the birth date of AI, which took place at the Dartmouth conference organized by John McCarthy and Marvin Minsky. Pitrat (2002a, p. 14) notices that the participants to this conference mainly presented ideas and projects, whereas Newell and Simon came with the Logic Theorist, a system that functioned and could solve problems that were nontrivial for humans.

8 Simon's Framework for Design: *The Sciences of the Artificial*

Simon's bibliography comprises nearly 1,000 titles, among which are some 700 papers published in journals in domains ranging from public management to the axiomatization of physical theories (*Bibliography of Herbert A. Simon-1930–1950's*). He published only some 10 papers directly concerned with design (Cagan, Kotovsky, & Simon, 2001; Kim, Javier-Lerch, & Simon, 1995; Simon, 1969/1999b, 1971/1975, 1973/1984, 1977b, 1980, 1987/1995c, 1997a). The number amounts to some 20 if one also includes publications dealing mostly with organizational design, but that do not handle with cognitive aspects.

The Sciences of the Artificial (Simon, 1969/1999b) is, however, one of Simon's seminal works and one of the definitely fundamental references exploited in cognitive analyses of design. The "sciences of design" are the core of these "sciences of the artificial" (or "artificial sciences," e.g., engineering, computer science, medicine, business, architecture, painting, the human and social sciences). Even if only two chapters of the book are dedicated specifically to the nature of design, this is the central issue of the entire book. Together with the paper on "The Structure of Ill-Structured Problems" (1973/1984), these are Simon's central publications in his work on design and are two of the main references in the rest of Part II of this book.

One may notice that "sciences of the artificial" may be a more appropriate appellation than "artificial sciences," which may also refer to the domains of artificial intelligence and artificial life (see, e.g., the presentation of the "Philosophy of the Artificial Sciences" course at the University of Wisconsin-Madison, retrieved May 20, 2006, from http://philosophy.wisc.edu/shapiro/Phil554/Phil554.htm).

The Sciences of the Artificial went into three, each time revised, editions. Its first, the 1969 edition, collected the three Compton lectures that Simon gave in 1968, as well as a 1962 paper, "The Architecture of Complexity: Hierarchic Systems." In the Compton lectures, Simon developed his thesis of the contingency of artificial phenomena that had been central to much of his research, at first in organization theory, next in economics and management science, and later in psychology. This 1969 edition introduced the chapter "The Science of Design: Creating the Artificial." In 1981, revised versions of the three Compton chapters were alternated with the three Gaither lectures delivered by Simon in 1980. This version also introduced a second chapter specifically on design, namely "Social Planning: Designing the Evolving Artifact." Taken together, the conclusions of the two design chapters constitute the main lines of a curriculum for design education formulated by Simon. In 1996, the third edition introduced a new chapter on complexity, "Alternative Views of Complexity." This edition is presented as a "revised" version in that it contains "new references" that record "the important advances that have been made since 1981 in cognitive psychology . . . and the science of design" (p. ix). Even if it presents numerous new references, the core of the text has not been amended. In his analysis of Simon's work, Carroll (2006) does not examine if the existing chapters changed between consecutive editions, but he notices an evolution in the nature of the new chapters. The addition of the chapter "Social Planning: Designing the Evolving Artifact" translates for him Simon "considering design as a social activity in several different senses" (p. 5).

In the present book, the page numbers for quotations from *The Sciences of the Artificial* come from the third printing of the third edition of the book (Simon, 1969/1999b).

From the first edition on, Simon considers the sciences of design as sciences in their own right. He sees them as distinct from natural science, which is traditionally considered as "the" "science." Yet, in a lecture given in 1987 (not included in *The Sciences of the Artificial*), Simon proposes to "compromise" on a perhaps less "pretentious" qualification, as he calls it, speaking of "the art and science of design" (Simon, 1987/1995c, p. 245). As Simon writes in the chapter titled "The Science of Design: Creating the Artificial" (in which engineering design is the reference), "historically and traditionally, it has been the task of the science disciplines to teach about natural things: How they are and how they work. It has been the task of engineering schools to teach about artificial things: How to make artifacts that have desired properties and how to design" (Simon, 1969/1999b, p. 111). Natural science is concerned with the necessary, with how things are, whereas design is concerned with the contingent, with how things might be (Simon, 1969/1999b, p. xii)—or *ought* to be.

Designers are "concerned with how things *ought* to be . . . in order to *attain goals* and to *function*" (Simon, 1969/1999b, pp. 4–5). Simon's thesis is indeed that "certain phenomena are 'artificial' in a very specific sense: They are as they are only because of a system's being molded, by goals or purposes, to the environment in which it lives" (Simon, 1969/1999b, p. xi). That is why symbol systems (or "information processing systems") are "almost the quintessential artifacts[:] Adaptivity to an environment is their whole *raison d'être*" (Simon, 1969/1999b, p. 22). "Artificial" indeed refers to human-made as opposed to natural. For Simon, our modern world is much more an artificial, that is, a human-made, than a natural world.

8.1.   SIMON'S ELABORATION OF AN SIP DESIGN THEORY

Two steps can be distinguished in Simon's elaboration of a cognitive design theory. The first one was taken together with Newell, to whom *Sciences of the Artificial* is being dedicated "in memory of a friendship." Jointly, the two researchers extended what has since been called the principles underlying the "symbolic information processing" approach to problem solving (Newell &

Simon, 1972)—or abridged the "symbolic processing" (Greeno & Moore, 1993, pp. 57–58), "symbolic" (Vera & Simon, 1993b, p. 10), or "information-processing" approach (Simon, 1978, p. 272), here abridged as the SIP approach. It is also frequently referred to—often by authors adopting a different approach—as the "rational problem-solving" (Dorst, 1997), "traditional," or "computational" view.

The SIP approach has been one of the main starting points of the "cognitivistic" perspective in cognitive science. In the early years of cognitive psychology, many authors embraced this paradigm as the fundamental schema for their investigation of cognitive activities. For some 20 years, it has been *the* theoretical reference for the cognitive analysis, not only of problem solving (G. A. Miller, Galanter, & Pribram, 1960; Reitman, 1965), but also of other types of activities: Concept learning (Bruner, Goodnow, & Austin, 1956), and verbal understanding and memory (Anderson, 1976, 1983; Le Ny, 1979, 1989a, 1989b). Together with various colleagues, Newell and Simon also used the approach to explore broader domains than the one analyzed in their famous *Human Problem Solving* (1972). They used it for their research into concept formation, verbal learning, and perception, but also administrative and organizational behavior, creativity and scientific discovery, and even music and emotion (for references, see Newell & Simon, 1972, p. 791, Note 1).

It was Simon alone—namely without Newell—who, subsequently, applied this paradigm to design (Simon, 1969/1999b, 1971/1975, 1973/1984, 1987/1995c). In these analyses of design, Simon identified and elaborated various characteristics of this specific problem-solving activity that have formed, for some 10 to 15 years, *the basis* of the approach adopted toward design by many, if not most, researchers in cognitive psychology and cognitive ergonomics who have been conducting research on design since the early 1980s.

With one exception (Okada & Simon, 1997), as far as we know, Simon was only concerned with individually conducted problem solving. This does not mean that he was a researcher who especially underestimated the importance of collective problem solving. In the 1960s and 1970s, few psychologists dealt with collectively conducted activities, analyzed from a cognitive viewpoint—there was, of course, research in social psychology, but these studies did not deal with cognitive aspects of problem solving.

Notice that the general reference for problem solving, the SIP model, was presented in 1972 (Newell & Simon, 1972), whereas the first edition of *Sciences of the Artificial* had already been published in 1969.

Simon's Analytical Approach to Design

Contrary to Simon's elaboration of a general theory of problem solving, which was based on experimental research, his work on design was analytical. With one or two exceptions (Kim et al., 1995), Simon indeed has not been involved in any empirical studies on design. This observation holds for "design" in a strict sense—such as Simon gave to the term. From the end of the 1950s on, Simon realized, in collaboration with various colleagues, a considerable body of research on scientific discovery, leading to two books (Langley, Simon, Bradshaw, & Zytkow, 1987; Simon, 1977a) and more than 40 papers (Cagan et al., 2001; Klahr & Simon, 2001; Kulkarni & Simon, 1988; Okada & Simon, 1997; Qin & Simon, 1990; Simon, 1977a, 1992c, 1992d, 2001b). Even if in our view, scientific discovery is based on the same cognitive activities and operations (and, of course, cognitive processes) as implemented in design, Simon nearly establishes no link with design (see, however, Cagan et al., 2001; see also our later section *Simon's More Nuanced Positions in Later Work*).

Reception of Simon's Design Framework

In 1964, Reitman adopted a representation for problem solving that could be formalized using the IPL-V information-processing language elaborated by Newell, Shaw, Simon, and other colleagues in the 1960s.[8] Reitman applied this problem-solving schema to the solving of what he qualified as "ill-defined" problems (see later discussion).

The architect Eastman (1969) was one of the first researchers to adopt the SIP framework for the analysis of design. He did so in what was at the time a particularly original study in the domain of empirical design research. He analyzed a protocol collected in a laboratory study concerning an architectural problem. Even if the problem was rather simple, his protocol study constitutes a reference in the domains of empirical studies of design, on the one hand, and of ill-defined problems, on the other.

There are also many authors who globally adopt Simon's framework, but propose more or less profound complements or modifications (Ö. Akin, 1986a, 1986b; Baykan, 1996; Goel, 1994; Goel & Pirolli, 1992; Hamel, 1995; Lebahar, 1983). Simon's ideas continue to be "a dominant force within the field," as noted by Roozenburg and Dorst (1999), who illustrate their claim by an analysis of the papers presented at the two first Design Thinking Research Symposia (DTRS) organized in Delft in 1992 and 1994 (Cross et al., 1992; Cross et al., 1996). They observe that "Simon was referred to more than anyone else: 31 direct references and goodness knows how many indirect ones in 32 papers" (p. 34, Note 3).

An explanatory hypothesis, which we have detailed in an analysis of 15 empirical design studies (Visser, 1992a, 1994a), is that the adoption by cognitive design researchers of rather strict SIP positions may be due to their data collection having been carried out in a laboratory or otherwise restricted context. An example is Goel (1995, p. 114) who observes and describes a quite orderly organization of the design process in different, consecutive stages. It should be noticed, however, that he has developed an innovative view with respect to a fundamental issue in cognitive modeling, that is, the status of representations. He did so around the notion of "sketch" (which we present later).

---

[8] IPL (information-processing language) was the first list-processing computer language.

From the end of the 1970s on, authors from various disciplines—psychology, sociology, ethnology, and anthropology—have been proposing other paradigms to the cognitive study of design (Bucciarelli, 1984, 1988; Schön, 1983, 1988, 1992; Rittel, 1972/1984, 1973/1984). Focusing on the so-called "situativity" approaches, we discuss them in Part III.

8.2.   SIMON'S DEFINITION OF DESIGN

Simon has provided various characterizations of design. Some general presentations translating his vision on the essential aspects of design are presented here. Other, more specific views are presented and discussed in later sections.

In *Sciences of the Artificial*, Simon (1969/1999b) proposes that "everyone designs who devises courses of action aimed at changing existing situations into preferred ones" (p. 111). Elsewhere in his book, he proposes a definition that may be related to the ASE paradigm:

> Design . . . means synthesis. It means conceiving[9] of objects, of processes, of ideas for accomplishing goals, and showing how these objects, processes, or ideas can be realized. Design is the complement of analysis—for analysis means understanding the properties and implications of an object, process, or idea that has already been conceived. (p. 246)

In a third definition (Simon, 1987/1995c), one may recognize the information-processing and artificial-intelligence modeling approach:

> Design is inherently computational—a matter of computing the implications of initial assumptions and combinations of them. An omniscient God has no need to design: The outcome is known before the process starts. To design is to gather information about what follows from what one has proposed or assumed. It is of interest only to creatures of limited information and limited computing power—creatures of bounded rationality like ourselves. (p. 247)

In a chapter written with Greeno (Greeno & Simon, 1988), design problems are analyzed as all other types of problems, that is, requiring search in a space, which, in the case of design, is a space that contains many possible arrangements of the problem materials, among which only one or a few satisfy the problem criterion. This view of design activity as *problem solving* in the standard information-processing sense of *search* in a problem space, is one of the characteristics of Simon's approach that is going to evoke much debate—for example, in our own approach to design.

The next five sections present five characteristics identified and elaborated by Simon, and that have formed the basis of cognitive design research by later researchers. They still occupy a central place in cognitive design studies, but we see in the next chapter that we consider that Simon misrepresented design with respect to several of these characteristics.

---

[9] In French, design is called "conception," and " designing "concevoir."

8.3. DESIGN
AS A TYPE OF COGNITIVE ACTIVITY
RATHER THAN A PROFESSIONAL STATUS

For ergonomists, Simon is an intellectual precursor when he states, in 1969, that "design" is not restricted to engineers. Simon (1969/1999b) considers that

> engineers are not the only professional designers. Everyone designs who devises courses of action aimed at changing existing situations into preferred ones. The intellectual activity that produces material artifacts is no different fundamentally from the one that prescribes remedies for a sick patient or the one that devises a new sales plan for a company or a social welfare policy for a state. . . . Engineering, medicine, business, education, law, architecture, and painting are "all centrally concerned with the process of design. (p. 111)

Nowadays, the view that design is a type of cognitive activity, not a professional status restricted to certain professionals, called "designers," is adopted by many, if not most, researchers in cognitive psychology and cognitive ergonomics of design. A difference with Simon's view on this issue is that they may not qualify, in the same way as Simon did, the domains of activity mentioned previously as "centrally concerned with the process of design."

Cognitive psychologists obviously agree on the approach of an activity in terms of the kind of mental processes and structures that are implemented, rather than in terms of the status of the person implementing the activity—be it their socioprofessional or a different status.

8.4. DESIGN AS SATISFICING

The concept of "satisficing" appears in one of Simon's first reports for the Rand Corporation, where he became a consultant in 1952 (as described by Pitrat, 2002a, p. 13). Simon also elaborated the corresponding idea in his first paper for a psychology journal (the *Psychological Review* in 1956).

Satisficing is a very general human approach, not specific to design. It has "its roots in the empirically based psychological theories . . . of aspiration levels" (Simon, 1978/1992b, p. 356). It was in the domain of economics that Simon first established its central role in human behavior. Simon (1969/1999b) describes his introduction of the notion as follows:

> Since there did not seem to be any word in English for decision methods that look for good or satisfactory solutions instead of optimal ones, some years ago I introduced the term "satisficing" to refer to such procedures. Now no one in his right mind will satisfice if he can equally well optimize; no one will settle for good or better if he can have best. But that is not the way the problem usually poses itself in actual design situations. In ["Economic Rationality: Adaptive Artifice," one of the 1980 chapters of *The Sciences of the Artificial*], I argued that in the real world

we usually do not have a choice between satisfactory and optimal solutions, for we only rarely have a method of finding the optimum. (pp. 119–120)

As an example of problem solving where satisficing will be used, Simon presents the "well-known combinatorial problem" of the "traveling salesman" (p. 120).

Satisficing thus is to "settle for the good enough" (Simon, 1971/1975, p. 1), to accept a satisfactory solution (Simon, 1987/1995c, p. 246), rather than to optimize, that is, to calculate the optimum value, or to choose the best solution among all possible solutions. When they satisfice, people decide without complete information. "Seldom will the goals and constraints be satisfied by only a single, unique design; and seldom will it be feasible to examine all possible designs to decide which one is, in some sense, optimal" (Simon, 1987/1995c, p. 246). Simon (2001a) writes:

> In the real world, problem solving seldom involves finding an optimum solution: Only rarely is this discoverable in spaces of real-life complexity, even with the largest computers. The expert searches until a solution is found that is good enough, that satisfices, that reaches what he or she thinks is a reasonable level of aspiration. (p. 207)

For people (compared to artificial cognitive systems), to satisfice is a *need*, and this need is mainly due to people's "bounded rationality" (Simon, 1982, 1997b). Bounded rationality is a typical human characteristic, which results from people's severely limited abilities with respect both to knowledge and information holding and accessing, and to computing power, that is, with respect to both people's "memory contents and their processes" (Simon, 2000, p. 25). It was for his idea of bounded rationality (1978/1992b) that Simon received the Nobel Prize in economics.

With respect to the selection criteria that may be used in choosing one, rather than calculating *the* solution, Simon, in one of his rare papers on design (Simon, 1971/1975), analyses "style"—in a very broad, not only aesthetic sense—as being one of the factors entering in "choosing any one of many satisfactory solutions."

Simon's approach to design as a satisficing activity has been widely adopted as central to design in cognitive design studies. Anticipating our discussion of Simon's position, we wish to note only the following. In the distinction between generation and evaluation activities in design (an opposition that Simon does not establish—not explicitly, in any case), Simon's satisficing view concerns evaluation rather than generation. For Simon, design alternatives are "found" (in a way that is not discussed). Yet, "once we have found a candidate we can ask: 'Does the alternative satisfy all the design criteria?'" (Simon, 1969/1999b, p. 121). Because it is rarely feasible to examine all possible designs in order to choose the optimal one, designs are then evaluated by comparing them in terms of "better" and "worse," more or less "acceptable," more or less satisfactory. Dealing with trade-offs among different possibilities plays a central role in this satisficing evaluation activity, a design characteristic that we further discuss when we present our approach to design.

8.5. DESIGN AS A REGULAR PROBLEM-SOLVING ACTIVITY: SIMON'S "NOTHING SPECIAL" POSITION

As briefly mentioned earlier, it was in their famous book *Human Problem Solving* from 1972, that Newell and Simon proposed to analyze problem solving as a symbolic information processing (SIP) activity.

For Simon, the central defining characteristic of the design activity is its problem-solving nature. The application of the SIP problem-solving paradigm to design was not made right from the start (i.e., in Simon et al.'s studies conducted during the 1960s, or in his work with Newell). Simon and colleagues were studying neither design, nor any other real-life activity, let alone a professional task. As noticed in the historical chapter, it is only in recent years that research on professional design situations has become relatively widespread. Newell and Simon note that their work is concerned with tasks that are short (half-hour), and that concern moderately difficult problems of a symbolic nature resolved by "intelligent adults" working individually. The elaboration, since the 1950s, of the 1972 SIP model had been based indeed on algebra-like puzzles (cryptarithmetic), chess, and symbolic logic problems—which we refer to as the "classical laboratory problems" in the rest of this book. Newell and Simon (1972) note that the thinking involved in, for example, "designing a house, discovering a new scientific law, . . . creating new music . . . are largely beyond the current state of the art," that is, as it was around 1970 (p. 7). Nevertheless, "they are in fact part of the same story [the authors] wish to tell" (p. 7). Moreover, when Simon started to tackle design problems and ill-structured problems, he maintained that no new and hitherto unknown concepts or techniques were necessary. According to Simon's "nothing special" position (presented in Klahr & Simon, 2001, p. 76, for scientific thinking), "no qualitatively new components" needed to be introduced in the classic general problem-solving mechanisms, in order to be able to handle design problems (Simon, 1973/1984, p. 197). No "special logic" was necessary (Simon, 1969/1999b, p. 115)—even if Simon "admits" that standard logic is to be adapted to the search for alternative solution elements (Simon, 1969/1999b, p. 124).

8.6.   DESIGN PROBLEMS:
"ILL-STRUCTURED" PROBLEMS?

A characteristic of design problems that nowadays is considered as one of its main specificities revealed by Simon, is their ill-defined, or in Simon's terms "ill-structured" character.[10]

We wish to call attention, however, to the fact that, even if Simon was one of the first authors to *discuss* design problems' structuredness in his famous paper "The Structure of Ill-Structured Problems" (1973/1984), he did not consider it a *specific* characteristic of design problems—and neither of another class of problems. What Simon underlines is that many problems often treated as well-structured are better regarded as ill-structured (1973/1984, p. 182).

Simon does not consider well- and ill-structuredness as categories in the classical, that is, pre-Roschian (Rosch, 1978) sense.[11] He starts his famous paper by discussing the impossibility to compose a formal definition of "well-structured problems." Instead he addresses "a list of requirements that have been proposed at one time or another as criteria a problem must satisfy in order to be regarded as well-structured" (Simon, 1973/1984, p. 182). "A problem may be regarded as well-structured to the extent that it has some or all of [these] characteristics" (Simon, 1973/1984, p. 183).

Reasoning from the example of designing a house, presented as a creative architectural-design problem, Simon (1973/1984) lists a number of characteristics that make this problem ill-structured. At the outset, when designers start to work on the design brief, and even after they have constructed a representation of the design project, their data does not fulfill the criteria of a well-structured problem. Several of the requirements referred to previously as "criteria a problem must satisfy in order to be regarded as well structured" are lacking. So are the following requirements:

- A definite criterion for testing any proposed solution, not to speak of a mechanizable process for applying the eventual criterion.

- One or more problem spaces in which one may represent the initial problem state, the goal state, and all other, intermediate states that may be reached, or considered, in the course of attempting a solution to the problem.

---

[10] For reasons exposed later, we adopt the terms "ill-defined" and "well-defined," except in discussions of the analyses made by Simon and other authors who use the terms "ill-structured" and "well-structured."

[11] The pre-Roschian view "in a nutshell . . . is the idea that categories and concepts are matters of logic; they are clearly bounded sets; something either is or is not in the category. It is in the category if it has certain defining features, and if it doesn't, then it's outside the category" (Scharmer, 1999).

- One or more problem spaces in which one may represent any knowledge that the problem solver can acquire about the problem.

- A possibility to define with complete accuracy the changes in the world that the design may bring about.

Simon thus does not consider "ill-structuredness" or "well-structuredness" an absolute problem characteristic. "The boundary between well-structured and ill-structured problems is vague, fluid and not susceptible to formalization" (Simon, 1973/1984, p. 181). "In fact, many kinds of problems often treated as well-structured are better regarded as ill-structured" (p. 182). Examples of such problems are chess playing and theorem proving, two types of problems on which the analyses underlying Newell and Simon's (1972) SIP model were based. Nevertheless, this does not lead Simon to conclude that for human problem solvers there are no well-structured problems, and that every real or even realistic problem is ill-structured. According to Simon, one must reconsider the nature of ill-structured problems. It is only at first analysis that "typically" "ill-structured" problems, such as creative design problems, may be considered ill structured. Reasoning on the example cited earlier (designing a house), Simon argues that even such a "typically ill-structured" problem rapidly acquires structure, due to a designer applying certain strategies, such as decomposition, and evoking from memory well-structured subproblems. "General problem-solving mechanisms that have shown themselves to be efficacious for handling large, albeit *apparently* well-structured domains should be extendable to ill-structured domains without any need for introducing qualitatively new components" (Simon, 1973/1984, p. 197; Simon's "nothing special" position).

With respect to the house design example, Simon states: "During any given short period of time, the architect will find himself working on a problem which, perhaps beginning in an ill-structured state, soon converts itself through evocation from memory into a well-structured problem" (Simon, 1973/1984, p. 190). Both chess-playing and architectural-design problems may be ill-structured "in the large," but they are well-structured "in the small" (Simon, 1973/1984, p. 191). "In the small" refers to the subproblems to be solved (in chess, playing one single move; in architecture, designing a heating system or another component of the building), "in the large" to the global problem (in chess, the complete game; in architecture, the entire project of the building).

In conclusion, if researchers refer to Simon for their characterization of design problems as "ill-structured" (or "ill-defined"), they misrepresent Simon's view. Simon was indeed one of the main authors at the origin of the *discussion* of design problems' structuredness, but for him, ill-structuredness was not a distinctive characteristic of design problems.

8.7.   SOLVING
ILL-STRUCTURED PROBLEMS
IN TWO CONSECUTIVE STAGES:
FIRST STRUCTURING AND THEN SOLVING

According to Simon, an ill-structured global problem—such as the design of a house—is thus rather smoothly amenable to one or more well-structured problems. It may seem ill-structured at first sight, but the designer structures it, and ends up with a well-structured problem! This holds in the same way for both typically ill-structured problems and seemingly perfectly well structured problems, such as algebra-like puzzles, the famous Tower of Hanoi, or symbolic-logic problems.

Solving ill-structured problems therefore proceeds through two consecutive stages, first structuring the ill-structured problem, and then solving the resulting well-structured problem(s).

Simon's formulates one remark with respect to the relative importance of this structuring activity. "There is merit to the claim that much problem-solving effort is directed at structuring problems, and only a fraction of it at solving problems once they are structured" (Simon, 1973/1984, p. 187). If one realizes how much time product-development projects spend in design, this "structuring" activity is then perhaps what characterizes the early conceptual-design stages (Sharpe, 1995), which are the core of design analyzed from a cognitive viewpoint.

In his comments and answers to questions formulated at the 1984 conference Sciences de l'Intelligence, Sciences de l'Artificiel [Sciences of the Intelligence, Sciences of the Artificial], Simon discusses whether "problem formulation" may be considered "problem solving." He concludes that the processes that lead from a problem situation to its solution are the same as those leading from a situation to the formulation of a problem (Simon, 1984/2002, p. 44). He refers to his paper on ill-structured problems (1973/1984) as a first approach to the analysis of these problem-formulation processes. Simon shows himself optimistic with respect to the possibility to model the activity involved in solving poorly structured problems. He bases this optimism on the following reasoning. To solve an ill-structured problem requires the formulation of goals—as do all goal-oriented activities such as problem solving. Even if the goals of an ill-structured problem are imprecise, they necessarily rely on criteria the person already possesses—be it implicitly. Therefore, Simon (1984/2002) states that, rather than to consider that goals are to be "formulated," it would be more appropriate to characterize the activity as "evocation" and "definition" of goals.

Many authors have adopted Simon's view on design problem solving in two stages. Ö. Akin (2004), for example, considers that "problem structuring is a prerequisite to problem solving: Problem structuring consists of a series of transformations converting an ill-structured problem to a well-structured one" (p. 43). Ö. Akin (1986a), however, also attributes a central role to "restructuring" design problems, something that "much simpler problems" hardly require "after the initial understanding phase" (p. 226). Examples of such "much simpler problems" are "certain puzzles, proofs and mathematical problems used in standard textbooks" (i.e., we add, the types of problems analyzed by Newell and Simon, 1972, in order to elaborate the SIP framework). Akin also observes that, in spite of vast differences between these much simpler problems and design problems, there are also many similarities. Both call, for example, for the same problem-solving stages of "problem understanding, resolution, and restructuring" (1986a, p. 226) (see also Goel's 1995 decomposition into stages).

9 Discussion of the Symbolic Information Processing Approach to Design

As may already be clear from several comments formulated during our presentation of Simon's view on design, we take a different stand with respect to designing as a problem-solving activity, to design problems as "ill-structured," and to the structuredness of the solving process applied to these problems.

We address these points in a series of criticisms addressed to the SIP approach, especially its underestimating certain aspects of design activity (such as the role of problem representation construction and of "nondeterministic leaps") and its overestimating other aspects (such as recognition, systematic problem decomposition, means-ends analysis, and search).

9.1. MISREPRESENTATIONS OF DESIGN BY THE CLASSICAL SYMBOLIC INFORMATION PROCESSING APPROACH

The discussion focuses on the approach that has been proposed by Simon as the researcher who has applied the classical SIP approach to design in *Sciences of the Artificial*. We also refer to Simon's collaborative work with Newell (1972) and Greeno (1988), and to work by other representatives of the SIP approach, such as Newell's own research (1969).

In general terms, our critique concerns SIP's too systematic, and thereby impoverished, approach to design: Simon represents design as much more orderly than actual design has been shown to be. The essence of Simon's view may well be applicable to small and rather simple, well-defined problems, and to their processing, but is unrealistic with respect to the typical ill-defined problems that professional designers will often meet.

After this discussion, the next chapter presents some more nuanced positions adopted by Simon and a hypothesis concerning the reasons underlying his view on design.

Underestimating the Specificity of Ill-Defined Problems and Their Solving

We adopt the terms "ill-defined" and "well-defined" as referring to both problem *structure* and the more or less well-specified character of other problem features. Newell (1969) has used the term "ill formed" in addition to "ill-structured."

The cognitive design research literature generally qualifies design problems as "ill-defined" (or "ill-structured") by reference to Simon (1973/1984). Yet, as we wrote already, Simon himself did not consider these problems as specifically "ill-structured" — chess and symbolic-logic problems were also. If the reference to Simon is nevertheless appropriate, it is because of his 1973 paper on the structuredness of design and other problems that initiated an important examination of this issue in design contexts.

According to Dasgupta (2003), "Simon's corpus of work was almost entirely in the realm of ill-structured problems . . . : Administrative decision making, economic decision making, human problem solving, scientific discovery, and design are all examples par excellence of the ill-structured" (p. 702). Dasgupta declares that "the realm of ill-structured problems" (p. 702) is to be taken in the sense in which Simon (1973/1984) himself has defined this concept. We consider that human problem solving in these various domains may pertain to "ill-structured problem solving," but not in Simon's sense! People who have to solve problems in administrative or economic decision making, designers and scientists who make discoveries, do not first "structure" their problems and then "solve" them — the two-stage process that Simon considers as the specificity of ill-structured problems solving.

With respect to the specificity of ill-defined problems and their solving, empirical observations on professional designers show that,

- even if well-defined (sub)problems exist, there are as well many other, ill-defined (sub)problems; and
- even if problems are well-defined at the level of the most concrete and detailed subproblems (at the level just before specification of implementation), many ill-defined problems will have to be solved before one attains that level.
- however, most important, there are not two separate phases in design: Structuring and solving. Design is no linear process (see the section *Stage Models and Process Models*). Of course, designers analyze problems, they decompose them, they interpret and reinterpret them, thus making them more manageable, easier to solve. Yet, it is all through the design process, until advanced project stages, that they occur, these so-called "structuring" activities that elaborate problem-solution related representations and procedures, through, for example, inference, exploration, simulation, reinterpretation, and analogical reasoning.

This last point has also been underlined by Mayer (1989) in his paper "Human Nonadversary Problem Solving," which presents a series of arguments that call into question several implicit premises of the SIP model. For Mayer, the "componential" character of problem solving is a premise of the SIP model that is denied by the reality of ill-defined problems. Mayer also denounces SIP's "atomization premise," that is, the principle that problems have clearly delimited "atoms": States and operators, in the case of SIP. For ill-defined problems, Mayer judges that "determining the atoms . . . is often the crucial part of problem solving rather than the starting point. Further . . . the atoms may change as they are combined during the problem solving process" (p. 54).

Even if the idea of designers proceeding in two consecutive stages is Simon's prevailing view in his 1973 paper, elsewhere he is occasionally less "extreme" and seems to consider most problem solving as switching between ill-structured problems (ISPs) and well-structured problems (WSPs) (Simon, 1973/1984, p. 197). "Each small phase of the activity appears to be quite well-structured, but the overall process meets none of the criteria we set down for WSPs" (p. 194). Problem solvers—designers—can be considered to be "faced at each moment with a well-structured problem, but one that changes from moment to moment" (p. 195). Simon does not analyze, however, or even mention, activities such as "reframing," "redefinition," or "restructuring" to which designers in that case should proceed with respect to their problem—a type of activity emphasized by his situativity opponents (see later discussion).

Even if we agree with Simon that "there is no real boundary" between well-defined and ill-defined problems, we think that it has, at least, a clearly heuristic function to distinguish "typically" ill-defined problems (such as design problems) from "typically" well-defined problems (such as the classical laboratory problems). This distinction makes it possible, for example, to predict and anticipate certain particularities of designers' activity when they are confronted with such problems (see later discussion).

Newell's (1969) argumentation as to this point seems more realistic. In "Heuristic Programming: Ill-Structured Problems," Newell starts by defending the position that Simon adopts in his 1973 paper, judging typically well-defined problems, such as logic-theorem proving, checker playing, and pattern recognition, as nevertheless ill-defined. He advances, however, as the reason for this position that, in the 1950s, algorithms for these tasks were either nonexistent "or were so immensely expensive as to preclude their use" (p. 366). Newell discusses Reitman's (1964) position that such problems may constitute a "small and particularly 'well-formed' subset" of ill-defined problems. According to Reitman, one may indeed lack well-specified algorithms for problems such as theorem proving, but the heuristic programs that, around the 1950s to 1960s, came to solve such problems, were themselves "otherwise quite precisely defined" (Newell, 1969, pp. 366–367). In addition, both the initial database from which the problem started and the test whereby one determined whether the problem had been solved, were well specified. However, where Reitman attempted to formulate "a positive characterization of problems by setting out the possible forms of uncertainty in the specification of the *problem*" (p. 367; the emphasis is ours), Newell tackles the question of ill-defined problems through an analysis of the problem-solving *activity* using problem-solving methods. Newell is conscious that, if he proceeds in this way, his approach "de-emphasizes some important aspects, such as the initial determination of an internal representation, its possible change, and the search for or construction of new methods (by other methods) in the course of problem solving" (Newell, 1969, p. 369). At the end of his paper, Newell comes back to these neglected "important aspects" and extensively discusses a series of "difficulties" encountered by his position.

Newell (1969) starts this *Difficulties* section asserting that heuristic-programming problem-solving methods are "only a part of problem solving." We listed earlier some of "the many [other] parts of problem solving" that he enumerated (recognition, evaluation, representation, information acquisition, method identification, method construction, executive construction, and representation construction). Newell considers if "the aspects of problem solving that permit a problem solver to deal with ill-structured problems reside in one (or more) of these parts, rather than in the methods" (p. 407).

With respect to the identification of a method, for example, he mentions that "much of the structuring of a problem takes place in creating the identification" (Newell, 1969, p. 409). For ill-defined problems, the difficulty of this creation resides in the identification from an unformalized environment to the problem statement of a precise, particular, formalized method.

Information acquisition, another example, might seem not particularly distinctive for solving ill-defined problems. For well-defined problems, the search for information is guided by a highly specific goal; whereas, in the case of ill-defined problems, information is to be acquired for ill- or underspecified use: It is "to be used at some later time in unforeseen ways" (Newell, 1969, p. 410). Information acquisition, therefore, could play "a central role in handling ill-structured problems" (p. 410).

The major difficulty of his position may be, according to Newell (1969), that it does not "come to grips directly with the nature of vague information. Typically, an ill-structured problem is full of vague information. This might almost be taken as a definition of such a problem, except that the term vague is itself vague" (p. 411). Newell assumes that "the notion of vague information is at the core of the feeling that ill-structured problems are essentially different from well-structured ones" (p. 412). Of course, a problem solver has a definite problem statement, but "all the vagueness exists in the indefinite set of problems that can be identified with the problem statement. . . . When a human problem solver has a problem he calls ill-structured, he does not seem to have definite expressions which refer to his vague information. Rather he has nothing definite at all. As an external observer we might form a definite expression describing the range (or probability distribution) of information that the subject has, but this 'meta' expression is not what the subject has that is this information" (pp. 411–412).

Simon (1973/1984) has also formulated some interesting observations with respect to the role of knowledge in relation to the ill-structured character of a problem—even if these observations remain isolated in his further "strict" approach to the structured character of problems and their solving. He states "there may be nothing other than the size of the knowledge base to distinguish ISPs from WSPs" (p. 197)—but he does not seem to consider that such a position makes problem structuredness into a relative problem characteristic.

Simon (1973/1984) advances,

> Any problem-solving process will appear ill-structured if the problem solver is a
> machine that has access to a very large long-term memory (an effectively infinite
> memory) of potentially relevant information, and/or access to a very large external
> memory that provides information about the actual real-world consequences of
> problem-solving actions. (p. 197)

We spot a presupposition underlying this prediction concerning the link between a cognitive system's amount of knowledge related to a problem, and its
view of the problem as ill-structured. Simon seems to presume that the impres-
sion of ill-structuredness augments with the size of the search space that is to be
traversed. We agree that the size of a search space may affect the effort to traverse it. Yet, we want to invert the direction of the dependence of ill-definedness
on the quantity of knowledge one possesses with respect to a task. We assume
that a task will appear relatively better specified if one has much potentially
relevant knowledge or information—even if its access may constitute a difficulty
(see the later subsection *Underestimating the Role of Nondeterministic Leaps.
Overestimating the Role of Recognition*).

Newell's (1969) approach to this issue may establish a bridge between
Simon's position and ours. In the same way as Simon, Newell claims that humans do not have strong, domain-specific "methods of unknown nature for dealing with ill-structured problems" (p. 407). In order to handle such problems,
humans use the same range of heuristics and other weak methods as they do for
solving rather well-defined problems. As presented earlier, Newell, however,
identifies that methods are "only a part of problem solving" and that there are
"other parts." Two of Newell's other "parts" that we consider essential are "representation" and "representation construction." Newell advances "the possibility
that only special representations can hold ill-structured problems" and that to
handle such problems is "to be able to work in such a representation." He has the
"suspicion" that "changes of representation [at a certain level] . . . might constitute a substantial part of problem solving" (p. 408). We will see the importance
of these activities ("parts") in our approach to design.

It should be noted that both Newell (1969) and Simon (1973/1984) discuss
the question of ill-structured problems in an AI context. Even if Simon extends
his discussion to human problem solving, it is Newell's position that seems—
relatively—more "human-realistic" or compatible with the way in which real
humans function.

Underestimating the Role
of Problem-Representation Construction

Simon has a rather inconclusive position with regard to the role of representation
in design. Even if he sometimes states that the construction of representations is
important in design, he generally does so in a final subsection or presents it as a

detail that "deserves further examination." It is clearly peripheral to an SIP model of design problem solving.

On several occasions, Simon starts with a very strict position concerning an issue that we consider critical, nuancing it little by little. With respect to representations, to the question if they are "found" or "constructed," Simon starts by writing that "every problem-solving effort must begin with creating a representation for the problem—a problem space in which the search for the solution can take place" (p. 108). He continues: "Of course, for most of the problems we encounter in our daily personal or professional lives, we simply retrieve from memory a representation that we have already stored and used on previous occasions" (p. 108). Sometimes, he admits, we have to adapt an existing representation. "Occasionally," he declares, a new representation, a new problem space, has to be discovered. "More often," one is midway between simply adapting a known representation and inventing a new representational system.

Elsewhere Simon uses the expression "discovering new representations" as if they have always been there, waiting to be found. Yet, he admits that the processes involved in this discovery are "a major missing link in our theories of thinking" and notices that they are "currently a major area of research in cognitive psychology and artificial intelligence" (Simon, 1969/1999b, p. 109). With respect to this, he refers to his study with Kaplan (1990), which indeed proposes some elements for such a theory.

When Simon suggests how a new representation may come into life, he states "focus of attention is the key to success—focusing on the particular features of the situation that are relevant to the problem" (Simon, 1969/1999b, pp. 108–109). This focus-of-attention idea constitutes a first step toward building "a theory of representation change" as Simon calls it.

In the chapter on social planning, Simon introduces at least six new topics for his design-education curriculum (from "bounded rationality" to "designing without final goals," 1969/1999b, p. 166), and he expands the "representation of design problems" topic. Simon notes that "the design tools relevant to these additional topics are in general less formal" than those described in relation to engineering design (p. 166). The domain of design thus seems to influence the nature of the cognitive processes and representations used in design problem solving (or "form" of design, as we qualify it; see our section *Design as a Specific Cognitive Activity, In Spite of Design Taking Also Different "Forms"*).

As Newell and Simon mention explicitly, their SIP theory of human problem solving does not deal with *changes* in problem representation—, in 1972 at least (pp. 90–91). In their 1972 vision, problem solving is search *in* a problem space that is constructed based on the problem requirements received at the start. The authors focus on the way in which, given a particular problem space, that is, a particular problem representation, a person tries to solve that problem. Among the problem solvers observed by Newell and Simon, only one paid specific attention to *choosing* a problem space.

The metaphor used to characterize solution processes adopted in order to solve typical classical laboratory problems is "searching through the set of possibilities." The corresponding metaphor proposed by Greeno and Simon (1988) for design problems is "narrowing down the set of possibilities" defined by a problem space. This is indeed the approach that Simon (1969/1999b) seems to adopt for design. At the beginning of the "Science of Design" chapter in *The Sciences of the Artificial*, he asserts that design requires making a rational choice among a set of given, "fixed alternatives," "computing the optimum" (pp. 114–119). In our opinion, proceeding in this way is probably an appropriate, if not "optimal" approach. Real designers, however, that is, professionals working on design projects, often do not consider "the set of possibilities." Identifying *the* set of possibilities would indeed be too heavy a task (cf. satisficing).

In a later section of his text, Simon (1969/1999b) observes that his initial position was not realistic: "In the real world," design alternatives are "not given in any constructive sense, but must be synthesized" (p. 121). They are even not given "in the quixotic sense that [the set of alternatives] is 'given' for the traveling-salesman problem" or given as the result of using a "formal but impracticable" algorithm (see also Newell's, 1969, position concerning the state of algorithms in the 1950s). Therefore, designers, rather than starting with a set of given alternatives among which they have to choose, are concerned with "*finding* alternatives" (the emphasis is ours). "Once [they] have found a candidate [they] can ask: 'Does the alternative satisfy all the design criteria?'" (Simon, 1969/1999b, pp. 119–121).

"But how about the process of *searching* for candidates?" (Simon, 1969/1999b, p. 121). Simon proposes means-ends analysis, which had been proposed in 1972 as the general strategy adopted for problem solving (Newell & Simon, 1972) (the plausibility of this strategy in design is discussed later).

Then, finally, near the end of the chapter, in a subsection entitled "Problem Solving as Change in Representation," Simon (1969/1999b) writes that "a deeper understanding of how representations are created and how they contribute to the solution of problems will become an essential component in the future theory of design." "Alternative representations for design problems" then becomes the final topic in Simon's curriculum for design education (pp. 132–134).

Based on their analysis of De Groot's observations of chess masters, Newell and Simon (1972) provided some preliminary, interesting ideas concerning "problem redefinition," which they did not further work out, however:

> Redefinition may come about (1) because a feature of the situation is noticed that has been overlooked earlier, (2) because initial expectations about the value of a move have been disappointed, or (3) because exploration of a move relates it to outcomes and goals different from those that suggested it. In these situations the dynamic analysis of moves appears more as an information-gathering process than simply as forward search through a branching space. (p. 762)

In the research conducted by Simon and colleagues on insight problems and scientific discovery, one may find some elements for the "future" theory of design in which an important role is attributed to representations, their creation, and modification. This research is described as representing "a major extension of the standard information processing theory of problem solving" (Kaplan & Simon, 1990, p. 376). According to Kaplan and Simon, their research brings the "good news" that "the same processes that are ordinarily used to search *within* a problem space can be used to search *for* a problem space (problem representation)" (p. 376). "But since previous evidence shows that subjects do not often switch their representations, [the authors] must also explain how the search for a new representation is initiated and under what conditions it has chances of success" (p. 376). We wish to underline that the restricted experimental nature of the problems and their solving process examined by Simon and colleagues may have contributed to this scarcity of "representation switching."

Underestimating the Role of Nondeterministic Leaps. Overestimating the Role of Recognition

Simon does not deny that intuition (Simon, 1987), insight (Kaplan & Simon, 1990), and inspiration (Simon, 1995a) may play a role in human activities. They are, however, not the result of any mysterious, unexplainable internal force. Schön (1983) concurs on this point with Simon:

> The idea of reflection on seeing-as suggests a direction of inquiry into processes which tend otherwise to be mystified and dismissed with the terms "intuition" or "creativity," and it suggests how these processes might be placed within the framework of reflective conversation with the situation which I have proposed as a partial account of the arts of engineering design and scientific investigation. (p. 187) (see also Cross, 2001a, pp. 88–89)

For Simon, the good news is not the reflective-activity framework, but his conviction that intuition, for example, is a "phenomenon which can be explained rather simply: Most intuitive leaps are acts of recognition" (Simon, 1969/1999b, p. 89). If someone is experienced in a domain, a problem in the domain is solved by simply "reminding." The solution is "retrieved by an action of recognition, the [problem] constituting a cue that [evokes] one or more appropriate productions" [12] (Simon, 1992a, p. 133). "As Pasteur once put it, 'Chance favors the prepared mind'" (p. 133). "We know that 'intuition' is our old friend 'recognition,' enabled by training and experience through which we acquire a great collection of familiar patterns that can be recognized when they appear in our problem situations" (Simon, 1999, p. vii).

So, rather than on search, intuition is based on recognition, which, in tasks in which perception may play an important role, such as in chess, often takes the form of pattern recognition. In the section on intuition referred to earlier, Simon (1969/1999b) is indeed concerned with chess grandmasters. In his discussion, Simon analyzes in terms of recognition, grandmasters' "intuitive leaps" from chessboard-position features to memory representations of these positions.

For such chess players, their representations of many positions indeed constitute familiar patterns, which may function as cues evoking "directly" the mem-

---

[12] In this analysis, Simon (1992a) reasons in terms of production systems.

ory representations of "interesting" moves. Design problems, however, may require "interesting" ideas that are not directly evoked by the representations constructed by the designer. They often depend on less "simple" "leaps": They require leaps *between* domains (Visser, 1991b). In this case, the number of possible memory candidates is huge and the chance that there already exist links between target and source representations is tiny.

Two examples of less than "simple" leaps. In our composite-structure design study (see Table 6.1), we observed designers performing leaps between, for example, the problem of designing "unfurling principles" for future antennas, and representations—first mental, and then externalized—of "umbrellas" and other "folding and unfolding" objects, such as "folding photo screens," "folding butterfly nets," and "folding sun hats." It seems implausible that the corresponding "intuition" relied on preexisting associations.

Neither do we suppose this to be the case for leaps between the problem of designing a support system for such an unfurling antenna, and "curtain rails," "train rails," and "railway catenaries." We suppose that these were analogical leaps that designers made between domains whose elements did bear no surface similarity.

In order to explain such analogical leaps, more complex structures or mechanisms may be required—not instead of recognition, but rather in addition to this basic mechanism.

Johnson-Laird (1989) proposes to analyze such "nondeterministic leaps" in terms of what he calls "profound analogies." He notices that there are, of course, forms of analogy "that can be retrieved by tractable procedures." However, "the processes underlying the discovery of profound analogies . . . cannot be guaranteed by any computationally tractable algorithm" (p. 313). Profound analogies involve "genuine human creativity," which Johnson-Laird claims is nondeterministic:

> The creation of a profound analogy is unlikely to depend on preexisting rules that establish mappings between the source and target domains. The innovation indeed depends on the invention of such rules. . . . No algorithm for searching the correct mapping can run in a realistic time beyond a certain number of links. (p. 327)[13]

---

[13] In a description of the EPAM program, built to simulate human rote verbal learning, Simon (1995a) presents a treelike discrimination net as the core of the system. He writes, "If the net has a branching factor of 4, then recognition of a net discriminating among a million stimuli could be achieved by performing about ten tests (4^{10} = 1,048,576). The EPAM model, its parameters calibrated from data in verbal learning experiments, can accomplish such a recognition in a tenth to a fifth of a second" (p. 942). This seems impressive, but we do not know how to estimate the distance (e.g., number of transitions) between representations corresponding to, for examples, an "unfurling principle" for antenna, and an "umbrella." In the context of his analysis of ill-structured problems, Simon states, on the other hand: "There is no way in which a large amount of information can be brought to bear upon these processes locally—that is over a short period of processing. If a large long-term memory is associated with a serial processor of this kind, then most of the contents of

"The more constraints that individuals can bring to the task—the more they know about potentially relevant domains—the more likely they are to find the illuminating source" (Johnson-Laird, 1989, pp. 329–330) (cf. our earlier remark about the relation between amount of knowledge and perceived degree of ill-definedness).

The critical step in creating a profound analogue consists in "seeing" one's target in such a way that its representation evokes an analogous source in memory! Once the corresponding, "appropriate" representation has been constructed, it is indeed "simply" an act of recognition.

Modestly, Johnson-Laird (1989) concludes that previous theorists had already emphasized many points made in his paper. "What, perhaps, has not been noted before are the computational consequences of the exercise of creativity in the discovery of original analogies" (p. 330).

In order to defend the straightforward, basic nature of ill-defined problem solving, Simon presented the essentially serial character of the problem-solving system, and its limited capacity confining it to work on only a few input elements and to produce only a small number of symbol structures as output (cf. Note 13 about the EPAM program). Comparable arguments can be advanced, in our opinion, in favor of Johnson-Laird's position. It may be hard to traverse with some acceptable degree of probability or in some acceptable time scale, the amount of associative links that are to be traversed between source and target structures, each in remote domains, in order to lead to a profound analogy. Neither forward (prospective) search (generation and test), nor backward (regressive, retrospective) search (means-ends analysis) is of much help in such a situation.

This point is related to another premise of the SIP model questioned by Mayer (1989, pp. 54–55): Its "mechanization" premise. Mayer notices that many problems, especially ill-defined ones such as design problems, require strategies that are "much less algorithmic [thus less 'mechanizable'] and more intuitive than means-ends analysis" (p. 55). It may also illustrate the SIP premise that Mayer qualifies as "concretization." In the SIP approach, each action results in a "movement from one concrete state to another. However . . . thinking sometimes occurs at a general or a functional level rather than at the level of specific problem states" (Mayer, 1989, p. 55).

Simon, however, does not consider the need—or perhaps the possibility—of such "nondeterministic" leaps. He admits that design takes place in a context in which "all potentially relevant information" is not fixed, not even present, right from the start. He concedes that "there is no need for this initial definition" (Simon, 1973/1984, p. 193). Yet, this is required for a problem to be considered well-defined. Simon reasons as if all problems—be they design or of another type—may be solved in one step. This step links either a process and a subprocess that the process calls via a subroutine structure, or two symbol structures, the

---

long-term memory will be irrelevant during any brief interval of processing" (Simon, 1973/1984, p. 192).

first one evoking the second one via this mechanism that "recognizes when certain information has become relevant" (Simon, 1973/1984, p. 193).

As noted previously, for Simon, "insightful" problem solving does not call for specific "creative" processes that are different from those observed in "standard" problem-solving settings. In this respect, Simon resembles his behaviorist ancestors, with whom his theories were in strong opposition. Behaviorist theorists also advanced that the same processes are involved in solving routine and nonroutine problems. It were the Gestalt theorists who judged that nonroutine problem solving involved "qualitatively different thinking" than routine problem solving (Mayer, 1989, pp. 51–52).

Overestimating the Role of Systematic Problem Decomposition

Another example of Simon adopting an approach to problem solving that cannot account for design in the usual professional design practice, is the role attributed to systematic problem decomposition, especially proceeding through balanced, stepwise refinement.

Referring back to Christopher Alexander's (1964) famous hierarchical decomposition method (see C. Alexander, 1963/1984), Simon posits problem decomposition as a powerful problem-simplification method, particularly useful in solving ill-defined problems. So do other researchers (see, e.g., Ö. Akin, 2004; Ö. Akin & Akin, 1997).

Note that in this respect, Simon's conception of human design greatly resembles the "transformation approach to design" frequently adopted in AI (Balzer, 1981; Barstow, 1984). A slight reformulation of four points of critique formulated by D. Brown and Chandrasekaran (1989, pp. 21–22) with regard to this provides the following comments:

- In some domains, the design constraints as stated may not be factorizable as anticipated, and there may be significant interactions between the designs that are chosen to meet parts of them.

- There is no guarantee that design can always proceed incrementally. In certain domains, it may be necessary to first design certain large subsystems before design can proceed within subsystems.

- Sometimes there may be knowledge directly available that "cuts a swath across the space," so that several constraints are realized simultaneously by a precompiled design that is recognized as applicable (implementation of such a way of proceeding has been observed by Guindon et al., 1987, in their empirical design study).

- In many domains, the problem is reformulated by decomposition, leading to the creation of a number of disjoint local spaces, each one corresponding to a subproblem.

With respect to D. Brown and Chandrasekaran's (1989) own theory of mechanical design, the authors notice the following.

In "relatively routine design," where one knows "effective problem decompositions," "compiled design plans for the component problems," and "actions to deal with failure of design solutions," "there is very little complex auxiliary problem-solving needed. It is not trivial, however, as plan selection is necessary, and complex backtracking can still take place. The design task is still too complex for simple algorithmic solutions or table look up.... [It still requires] knowledge-based problem-solving" (p. 34).

In "open-ended creative design," where there is no "storehouse" of effective decompositions and of design plans for subproblems, if ever problem decomposition knowledge is available, most of the effort will be in searching for potentially useful decompositions (p. 33).

Both empirical studies and theoretical analyses show that, even if, in theory, decomposition is a strong and useful method, their application by designers rarely follows a simple procedure executable in a straightforward manner. In other than relatively routine design projects, designers certainly decompose, but often not in a systematic way.

We already pointed out that the subproblems resulting from decomposing a complex global design problem are seldom independent (D. Brown & Chandrasekaran, 1989). Simon notes himself that the interdependencies among the resulting subproblems "are likely to be neglected or underemphasized." "Such unwanted side effects accompany all design processes that are as complex as the architectural ones we are considering" (Simon, 1973/1984, p. 191). Therefore, many problem-solving activities will still be required in order to handle the interdependencies between subproblems, for example, to articulate and integrate their different solutions. The integration of solution components into one global solution will often come up against their nonadditivity, their nonindependence, and their interaction.

In addition, one and the same design component often can be decomposed in different constituents (Reitman, 1964, p. 296). Ratcliff and Siddiqi (1985) observe data-driven and goal-driven problem decomposition in program design. Decomposition can result in design constituents, but also in successive processing steps—something Jeffries et al. (1981) observed as a characteristic that distinguished beginning and advanced students.

A third complication concerns the application of "transformational formulas." Decomposition can indeed be implemented using decompositional rules and methods from a repertoire possessed by professionals experienced in the domain. In this line, Simon argues that design acquires structure through the application of "formulas" such as: "House" transforms to "general floor plan plus structure," then "structure" transforms to "support plus roofing plus heating plus utilities," and so on (Simon, 1973/1984, p. 190). For this approach in terms of transformational formulas, Simon refers to Reitman (1964). Reitman notes, however, that this analogy, imported from structural linguistics, is useful, but "must not be carried too far." There is no fixed limit on the number of sources that a designer may consult for transformational formulas. Besides, new formulas may be developed. Furthermore, based on unforeseeable decisions and information sources, decomposition may follow another than the specified order.

Finally, as noted by Goel (1995), decomposition may also result in modules that are "leaky."

The major basis for designers' decomposition is often not the structure of their problem task, but their experience (Goel & Pirolli, 1992). The meaning of this observation is not straightforward. In a study of social-sciences problem solving (particularly in the political-science domain), Voss, Greene, Post, and Penner (1983) observed two modes of subproblem generation. Subproblems resulted from deliberate decomposition, or were "encountered" when problem solvers explored the implications of proposed solutions. The two problem-generation modes were distributed over experts and novices in a way that might be considered counterintuitive. Experts primarily used the "encountering" mode, whereas novices typically proceeded by problem decomposition. Based on a comparative analysis of four groups of problem solvers,[14] the authors concluded that the use of the problem-decomposition strategy might be analyzed as the falling back upon more general problem-solving strategies when one is confronted with a problem outside one's specialty—that is, when one is, or becomes again, a novice.

This conclusion concurs with our observations concerning the "counterintuitive data from empirical studies" regarding the differential use of opportunistic strategies by novices and experts (Visser, 1991c) (see later discussion).

A characteristic of systematic decomposition that is not necessarily a drawback of the method is that it generally leads to standard, routine solutions. Indeed, decompositions are based on a priori categories, leading to the attribution of "default" values to the different components (see also Carroll, 2000, p. 27). This may be what a designer searches for, but will not be systematically the case. It strengthens Voss et al.'s (1983) interpretation.

Overestimating the Role of Search

Newell and Simon (1972) postulate that "problem solving takes place by search in a problem space"—an "odyssey" through the problem space (Simon, 1978, p. 276) (cf. also Gilhooly's, 1989, presentation of Duncker's Gestalt definition of "problem solving" quoted earlier). In their concluding theory section of *Human Problem Solving*, Newell and Simon (1972) postulate that this problem-space search principle is "a major invariant of problem solving behavior that holds across tasks and subjects" (p. 809). In his conclusion of the chapter on "The Psychology of Thinking"[15] in *The Sciences of the Artificial*, Simon states that

---

[14] These were (a) experts in the domain, (b) novices in the domain, (c) graduate students that formed a "transition" between these experts and novices, and (d) "nonexpert experts," namely advanced graduate students and faculty members in political science, but whose field was not the particular problem domain, that is, the Soviet Union.

[15] This chapter from 1969 was maintained as such in the 1996 third revised edition.

"the theory of design is that general theory of search," that is, "the general theory of search through large combinatorial spaces on the outer side [of the human brain, that is] the side of the task environment" (Simon, 1969/1999b, p. 83).

For Newell and Simon (1972), this does not mean, however, that "all behavior *relevant* to problem solving" is search. For example, "defining the situation," which may lead to "problem redefinition," "contrasts sharply" with search and thus is "an important type of information-processing to understand" (p. 761). Neither Newell and Simon, nor Simon (1969/1999b) did, however, develop this topic.

"Problem solving as search" models focus on traversal of spaces comprising the solutions that are *possible*—in theory or in the designer's representation. Such solutions are already *there* and are "simply" to be located in the state space or problem space (see also Gero, 1998a, p. 167). In design, problem representations (problem spaces) are, however, often *not readily available* through memory retrieval: They are to be *constructed* (cf. also design as exploration, Gero, 1998a, 1998b; Logan & Smithers, 1993; Navinchandra, 1991).

Overestimating the Role of Means-Ends Analysis

The main problem-solving method proposed by Newell and Simon (1972) as identified in their studies on classical laboratory problems, is means-end analysis. Given Simon's "nothing special" view with respect to ill-defined problems, this method is apparently also supposed to be used in order to solve design problems. In *The Sciences of the Artificial*, Simon indeed explicitly proposes means-ends analysis in response to the question: "But how about the process of *searching* for candidates?" (Simon, 1969/1999b, p. 121).

In our view, however, this heuristic will generally be inappropriate for solving design problems. Indeed, in order to use it, one must be "able to represent differences between the desired and the present." "The desired" is the problem's goal state; "the present" is the problem's current state (the initial problem state or an intermediate state). If differences are found, a mechanism will be brought into play that is equivalent to GPS ("equivalent," not identical, because "GPS operates . . . on formally presented problems, not on an external real world") (Simon, 1973/1984, p. 184). GPS is the computer problem-solving program designed to "model some of the main features of human problem solving." It uses a "table of connections, which associates with each kind of detectable difference those actions that are relevant to reducing the difference." When a difference is detected, GPS "searches selectively through a (possibly large) environment in order to discover and assemble sequences of actions that will lead from a given situation to a desired situation" (Simon, 1969/1999b, pp. 121–123).

For design problems, however, it will generally be difficult to establish the differences between these two states—that is, if ever "the" goal state has already been specified. And even if both states are specified, completely and definitively specified, often it will still be difficult, if not impossible, to compare them, because they will often be specified at different levels: Typically, the current state of the problem will be at a more "concrete" level than its goal state, which generally is, at best, at an abstract level.

Therefore, it seems to us that means-end analysis will seldom be an appropriate heuristic for handling design problems.

9.2.   SIMON'S MORE NUANCED
POSITIONS
IN LATER WORK

Simon's position with respect to design is not consistent over time. Understandably, more "nuanced" positions appear in more recent texts, mostly on scientific discovery and creative thinking (Cagan et al., 2001; Kaplan & Simon, 1990; Klahr & Simon, 2001; Kulkarni & Simon, 1988; Simon, 1992c, 1992d, 1995a, 2001a, 2001b). Among his two or three rare publications that explicitly deal with design, there is, however, also an *early* paper in which Simon presents a more subtle position concerning the ill-defined character of design problems. The paper—"Problem Forming, Problem Finding, and Problem Solving in Design"—published in 1995, was based on a conference presented in 1987, that is, several years after Simon's famous "The Structure of Ill-Structured Problems" (1973/1984). In this 1987/1995c paper, Simon, for example, states:

> We cannot really regard the goals of design as given any more than we regard [the initial states] as given. A design process begins with some criteria and some possibilities (or primitives out of which alternatives can be constructed). As the process goes forward, new criteria and new possibilities are continually being evoked from the sources we have identified. (p. 253)

In one of the "questions to Simon" that we formulated in our contribution to the International Conference in honour of Herbert Simon "The Sciences of Design. The Scientific Challenge for the 21st Century" (Visser, 2002c), we expressed our surprise that Simon's more "flexible" position expressed in 1987 (or perhaps only in 1995) was not reflected in the 1996 version of his "Science of Design" chapter.

In Simon's 1973 paper on ill-structured problems, which was largely quoted and discussed earlier, there is also an interesting assertion concerning the "illusory" character of problem structuredness and its "elusiveness." "Definiteness of problem structure is largely an illusion that arises when we systematically confound the idealized problem that is presented to an idealized (and unlimitedly powerful) problem solver with the actual problem that is to be attacked by a problem solver with limited (even if large) computational capacities" (Simon, 1973/1984, p. 186). This claim was, however, isolated in the further context of the paper. It did not lead Simon, for example, to conclude that, for human problem solvers anyhow, all real problems are ill-defined.

Scientific Discovery and "Inventive" Design

Most more-nuanced positions thus make their appearance in texts concerning scientific discovery and creative thinking. In a review on creativity, Simon (2001a) points out that "creative design" is not a "common-place" problem-solving activity (pp. 205, 206). In a 2001 state of the art paper based on findings by psychologists and other researchers concerning the process of scientific discovery, Klahr and Simon (2001) affirm:

> Initial [problem] state, goal state, operators, and constraints can each be more or less well-defined. For example, one could have a well-defined initial state and an ill-defined goal state and set of operators (e.g., "make something pretty" with these material and tools), or an ill-defined initial state and a well-defined final state (e.g., find a vaccine against HIV). But well-definedness depends on the familiarity of the problem-space elements, and this, in turn, depends on an interaction between the problem and the problem solver. (p. 76)

Another recent, particularly interesting paper compares scientific discovery and inventive engineering design, with respect to their cognitive and computational similarities (Cagan et al., 2001). Cagan et al. defend the "nothing special" position with respect to insight and scientific discovery problems—as Simon did regarding design problems and ill-structured problems:

> Perhaps the most important feature of the theory [of discovery] is that it is built around the same two processes that have proved central to the general theory of expert human problem solving: The process of recognizing familiar patterns in the situations that are presented, and the selective (heuristic) search of a problem space. Basic to these are the processes of formulating a problem space and of representing available information about the problem, as well as new information acquired through observation and experimentation, in that problem space. (p. 450)

In these recent papers, one may find other statements concerning the important role of representation in problem solving. "Representational changes" play a role in scientific discovery in that they enable scientists "to replace [an] entrenched idea" with new ideas (Klahr & Simon, 2001, p. 77; see also Kaplan & Simon, 1990).

In a paper on the importance of "curiosity" for "discovery" (Simon, 2001b; see also Simon, 2001a), Simon posits that—at least in certain cases—"creative" ideas or hypotheses come from "noticing some phenomena" and "recognizing some things about them" (Simon, 2001b, p. 10). Even if Simon does not insist on the role of knowledge, he mentions that something that we would call "prior knowledge" governs the occurrence of these processes: "We mainly notice things that are unusual or surprising in their current surroundings," that is, "when we have expectations that are violated" (Simon, 2001b, p. 10). Different types of data sources "all provide evidence that the scientist's reaction to phenomena . . . that are surprising can lead to generating and testing new theories" (Klahr & Simon, 2001, p. 78). For Klahr and Simon, scientific discovery is a type of problem solving using both weak methods that are applicable in all disciplines and strong methods that are mainly domain-specific. . . . Recognition processes, evoked by familiar patterns in phenomena, access knowledge and strong methods in memory, linking the weak methods to the domain-specific mechanisms. (p. 78)

The role that Simon attributes to *surprise* in discovery may be related to the description that Schön and Wiggins (1992) provide of "seeing-moving-seeing" design actions as possibly leading to the "discovery of unintended consequences" of these actions. Schön (1992) also emphasizes the role of recognition in the discoveries that he identifies as crucial in the development of design (e.g., p. 7).

Other authors have discussed the relation between design and science, and between design and discovery. In an analysis of the structure of design processes, Dasgupta (1989) considers design problem solving a special instance of scientific discovery.

In their comparison between design and discovery, Cagan et al. (2001) advance that "[highly creative] design activities are often labeled *invention*." The authors explain how the "seemingly disparate activities" of discovery and invention are surprisingly similar (p. 442). The authors' "major conclusion" of this comparison is that "at a deep level, the cognitive and computational processes that accomplish [design and discovery] are virtually identical" (p. 463). Their "real similarity" is made up by "the underlying cognitive activities based on problem solving, pattern recognition, analogical reasoning, and other cognitive knowledge retrieval mechanisms" (pp. 452–453).

"The fundamental difference" between the two is "the *goal* of the process: Scientific explanation versus creation of a new artifact. . . . Design starts with a desired function and tries to synthesize a device that produces that function. Science starts with an existing function and tries to synthesize a mechanism that can plausibly accomplish or account for that function" (Cagan et al., 2001, pp. 452–453). However, "their cognitive models of search, once the problems are defined, are the same" (p. 455).

An implementation of these ideas can be found in A-Design, a program proposed by Campbell, Cagan, and Kotovsky (1999, 2000) in order to automate the invention of electromechanical artifacts. "At a high level the A-Design program is a model of (human) group activity" (Cagan et al., 2001, p. 459). The observation that in human creative design, "random directions are chosen and pursued if seemingly beneficial or else reversed if seemingly inferior" motivates the search strategy of "simulated annealing" in A-Design and similar programs (Cagan et al., 2001, p. 461).

Still other quite nuanced claims appear in these and other recent texts, but they remain isolated in the further context of Simon's work. All these claims largely open up to positions attributing more influence to representational activities—as we are proposing in this book! Neither Simon, nor his collaborators, however, elaborated on these ideas! Surprisingly, Simon did not integrate these positions in the later editions of *Sciences of the Artificial*, his reference work on design—and one of his reference publications in general!

9.3. A POSSIBLE EXPLANATION OF SIMON'S APPROACH TO DESIGN

We have formulated a hypothesis in order to explain—at least, in part—Simon's view of design. It is based on Simon considering differently both

- cognitive activities in economics and in design, and
- design activities in engineering and in social planning.

As regards the first point, with respect to design, Simon seems to underestimate human cognitive limitations—as illustrated previously in great detail. In contrast, Simon's position concerning cognitive activity displayed in economics is much subtler than that concerning design thinking. Analyzing economic theories, Simon is very sensitive to the way in which they idealize human rationality and neglect its limits. As noticed already, this completely new approach in terms of "bounded rationality" bestowed him with the Nobel Prize in economics.

In the domain of economics, Simon ascribes the idealization of the human person to economic theories (especially neo-classical ones) directing "their attention primarily to the external environment of human thought, to decisions that are optimal for realizing the adaptive system's goals." These decisions "would [indeed] be substantively rational in the circumstances defined by the outer environment" (Simon, 1969/1999b, p. 23). In Simon's view, actual human economic behavior "illustrates well how outer and inner environment interact and, in particular, how an intelligent system's adjustment to its outer environment . . . is limited by its ability, through knowledge and computation, to discover appropriate adaptive behavior" (Simon, 1969/1999b, p. 25). In his 1981 chapter "Economic Rationality: Adaptive Artifice," Simon insists on the often satisficing nature of the economic actor's decisions. "Because real-world optimization, with or without computers, is impossible, the real economic actor is in fact a satisficer, a person who accepts 'good enough' alternatives, not because less is preferred to more but because there is no choice" (Simon, 1969/1999b, p. 25).

Under the title, "Satisficing Man Revisited," Reitman (1964) quotes a description that Simon (1959, pp. 272–273, quoted by Reitman, 1964, p. 313) gives of some differences between the traditional approach to choice behavior in economic theory and his own conception of satisficing:

> The classical theory is a theory of a man choosing among fixed and known alternatives, to each of which are attached known consequences. But when perception and cognition intervene between the decision maker and his objective environment, this model no longer proves adequate. We need a description of the choice process that recognizes that alternatives are not given but must be sought; and a description that takes into account the arduous task of determining what consequences will fol-

low on each alternative. . . . As every mathematician knows, it is one thing to have a set of differential equations, and another thing to have their solutions. Yet the solutions are logically implied by the equations they are "all there," if we only knew how to get to them! By the same token, there are hosts of inferences that *might* be drawn from the information stored in the brain that are not in fact drawn. The consequences implied by information in the memory become known only through active information-processing, and hence through active selection of particular problem-solving paths from the myriad that might have been followed.

In the domain of management science, Simon notices that more or less "rational" methods, namely methods from operations research to AI, "have been applied mainly to business decisions at the middle level of management. A vast range of top management decisions (e.g., strategic decisions about investment, R&D, specialization and diversification, recruitment, development, and retention of managerial talent) is still mostly handled traditionally, that is, by experienced executives' exercise of judgment" (Simon, 1969/1999b, p. 28). He adds: "As we will see in [the chapters on the psychology of cognition], so-called 'judgment' turns out to be mainly a non-numerical heuristic search that draws upon information stored in large expert memories" (p. 28).

As regards the second point—design activities in different domains—, it looks as if Simon's approach to design seems to be based on his view of what he presents implicitly as the prototype of design: Engineering design (and, to a lesser degree, architectural design). In his discussion of social planning, Simon (1969/1999b) states that in this form of design "representation problems take on new dimensions" relative to the "relatively well-structured, middle-sized tasks" of engineering and architectural design (p. 141). Simon assumes that, for "real-world problems of [the] complexity" of social planning, designers may refer to "weaker" criteria than for standard design. With respect to the social-planning problem of regulating automobile-emission standards, for example, Simon (1969/1999b) writes:

> One may regard "defensibility" as a weak standard for a decision on a matter as consequential as automobile emissions. But it is probably the strictest standard we can generally satisfy with real-world problems of this complexity. [Especially in situations of this kind] an appropriate representation of the problem may be essential to organizing efforts toward solution and to achieving some kind of clarity about how proposed solutions are to be judged. Numbers are not the name of this game but rather representational structures that permit functional reasoning, however qualitative it may be. (p. 146)

In the section *Designing Without Final Goals* of this "Social Planning" chapter, Simon affirms that processes such as "search guided by only the most general heuristics of 'interestingness' or novelty" may provide "the most suitable model of the social design process" (Simon, 1969/1999b, p. 162). He observes that this kind of search also provides the mechanism for scientific discovery.

It seems thus that only when he discusses economic and social problems (and perhaps scientific discovery) that Simon seriously considers human's bounded rationality and takes into account the role of what he calls "representations without numbers," generative constraints such as "interestingness" or "novelty," and critical constraints such as the "defensibility" of a decision.

With respect to the sources of social design problems' greater "complexity," Simon suggests that differences of at least three types may be involved: Problems' degree of structuredness, their size, and the nature of their object. We come back to this point in our section *Design as a Specific Cognitive Activity, In Spite of Design Taking Also Different "Forms"*.

Thus, Simon's approach of design could be less mysterious if one might suppose that he considers

- *routine* engineering and architectural design as standard design.
- social planning (and, maybe, also inventive engineering design, see previous discussion[16]) as *a form* of design that is *radically different* from standard design.

9.4.   CONCLUSION CONCERNING THE SIP APPROACH

We discussed earlier six aspects of design misrepresented in the classical symbolic information processing approach. They constitute, depending on one's optimism and one's confidence in the capacities of adaptation by the SIP approach, either a series of nuances with respect to this approach that might guide its modification, or such severe critiques that one considers the approach to be indefensible.

We have discussed, as it were, two "types" of positions existing in parallel adopted by Simon. First, we saw his rather "strict" SIP position concerning design as regular problem solving that deserves, and needs, "nothing special." Afterward, this chapter has introduced a more nuanced position, defended, on the one hand, in papers concerning discovery and creative thinking, and on the other hand, in the paper with Cagan and Kotovsky concerning inventive engineering design. We may relate these different positions defended in parallel by Simon to what we have noticed previously concerning Simon's divergent positions with respect to social planning compared to design thinking in general. We indeed observed that Simon considered "standard" engineering and architectural design as a "simple" problem-solving activity similar to that required by the classical laboratory problems. Social planning, at the other hand, was viewed by Simon as a "special" design activity in which, for example, representation issues play an important role.

In his *Sciences of the Artificial* (1969/1999b), Simon proposed a new vision on science, considering the sciences of the artificial, these sciences of design, as

---

[16] Yet, in the aforementioned paper on inventive engineering design, Cagan et al. (2001) claim that "invention is not essentially different from other types of design activities" (p. 451). The "nothing special" position traverses all Simon's work!

sciences in their own right, distinct from the natural sciences, which traditionally are considered "the" "science." Simon's conception of design translates, however, also a very conventional view: Management and administrative decisions, that is, actions in the social domain, are fuzzy, imprecise, and require a qualitative approach, whereas engineering and other "typical" design decisions are precise, exact, and can be handled with quantitative methods!

It is hoped that this part has provided to the researchers who are interested in design or in Simon, an original and useful contribution among the increasing number of "Simon Studies" (Dasgupta, 2003, p. 703) that are dedicated to this prodigious researcher, especially since his death in 2001 (see e.g., Dasgupta, 2003; Forest, Méhier, & Micaëlli, 2005; Perrin, 2002; Pitrat, 2002b).

Carroll (2006), in an "elaboration" on Herbert Simon's "Science of Design," revisits Simon's analysis of the ant crossing a beach (Simon, 1969/1999b, pp. 51–52):

> The lesson Simon draws from the parable of the ant is that the apparent complexity of organisms, including human beings, derives largely from the complexity of the environment within which they act. I found this lesson riveting, but incomplete. . . .
> I wrote a letter to Simon asking whether he agreed with a further conclusion regarding the ant; namely that, in order to understand the underlying structures of human cognition, one would have to describe in detail the tasks, the technology, the social conventions, and all other environmental features that contribute to human performance. It was a great thrill to me when he agreed with this point. However, in retrospect after two decades, it's a pretty obvious point. The parable of the ant, however vivid and stimulating, actually obscures the point by placing the ant on a beach, a somewhat randomly structured, and therefore arbitrarily complex, environment. A better example might be an ant navigating the corridors of an ant colony.
> I am ready to suggest another elaboration of the ant: We must talk to the ant, work shoulder to shoulder with the ant, and walk a mile in the ant's shoes if we really want to understand the beach and the ant's trajectory as it crosses the beach. (p. 17)

Carroll's approach to this communication is participatory design. We see other, not at all contradictory, approaches in Part III of this book, which presents alternative frameworks to design, focusing on situativity (SIT), and in Part IV, which presents our own position. Afterward, in Part IV, we confront SIP and SIT.

III

Modern Views on Design. The Situativity Approach (SIT)

The classical approach to cognition as it has been applied to design (discussed in the Part II) has been severely criticized for some 20 years now. Some of these criticisms are based, in our view, on terminological confusions, such as those referred to earlier concerning "problem" and "problem solving." There are other criticisms that we consider justified and that, for an important part, are related to criticisms formulated here. This book is, as far as we know, one of the first to outline and discuss in somewhat detail these criticisms in a confrontation with the SIP positions (see Part IV). There has been of course Dorst's doctoral thesis (1997) and papers taking up Dorst's positions (e.g., Roozenburg & Dorst, 1999), which also compared the two paradigms. We introduce nevertheless other dimensions in our analysis of both families of approaches, so that both comparisons may complement each other.

In his Introduction to *Bringing Design to Software*, Winograd (1996), under the heading "What Is Design?," qualifies design through the titles of several sections, in the following terms:

- Design is conscious.
- Design keeps human concerns in the center.
- Design is a conversation with materials.
- Design is creative.
- Design is communication.
- Design has social consequences.
- Design is a social activity.

That is why he considers that one needs to adopt a "situated" perspective to design. This "concern for the situated nature of design—a sensitivity to the human context in all its richness and variety" is "what is common to all the authors" of his book (p. xxv).

Depending on authors and on disciplines, "situated" has different meanings. According to G. J. Smith and Gero (2005), "a common AI understanding . . . equates ['situated'] with 'embodied'" (p. 535).

Greeno and Moore (1993) have proposed the term "situativity" to cover approaches in terms of both "situated cognition" and "situated action." They consider that "the phrase *situated cognition* often is interpreted, understandably, as meaning a kind of cognition that is different from cognition that is not situated" whereas the authors consider situativity "a general characteristic of cognition" (p. 50). According to these authors, the situativity perspective claims, "cognitive activities should be understood primarily as interactions between agents and physical systems and with other people" (p. 49). It may be noticed that Greeno was one of Simon's collaborators before he embraced the situative perspective.

Outline of this part. This part discusses views on design activities proposed as alternatives to the SIP approach, focusing on viewpoints that are characterized as "situated." It distinguishes two phases in these approaches, the second one reviewing and developing the first one on certain points. The early period is mainly represented by Donald Schön. A central author in the second one, still in progression, is Kees Dorst in collaboration with several colleagues. First, however, other alternative and related positions are discussed with respect to the situated approaches.

10 Situativity and Other Alternatives to the SIP Approach

From the 1980s on, when cognitive design studies began to flourish, researchers have formulated, with respect to the SIP approach, alternative views on human design in its actual accomplishment.

10.1. ALTERNATIVE APPROACHES TO DESIGN

In this context of renaissance, researchers have shown renewed interest in older theories, especially in "activity theory" (Leontiev) and Piaget's views. In this book, we focus on the situative perspective (SIT).

We view situativity as a general framework that may embrace viewpoints on design that are more specific. We consider as instances of situativity, for example, "reflective practice," "reflection-in-action," and "knowing-in-action," on which Schön focuses.

Authors frequently referred to in the context of situativity theory and who are among the forerunners of the underlying approaches are Winograd and Flores (1986), Suchman (1987), Agre and Chapman (1987), Lave (1988), J. S. Brown, Collins, and Duguid (1989), and Clancey (1991). Many sympathizers of the situativity approach are working in the teaching and learning community[17] (e.g., Greeno, 1997, 1998; Greeno & Moore, 1993; Lave & Wenger, 1991; see also the *Educational Researcher* debate presented later).

Another approach related to situativity that attracts many authors who study collaboratively conducted activities, analyses situations in terms of "distributed cognition" (Hollan, Hutchins, & Kirsh, 2000; Hutchins, 1995, 1996). It is not further discussed in this book. First, it has not been applied in particular to designing. Second and more important, we consider that this approach, which considers human and artificial agents as symmetrical partners in a cognitive system, does not establish an appropriate differentiation between human and artificial cognition.

Roozenburg and Dorst (1999) present Coyne as an exponent of "design theorists who approach design from 'post-modern,' that is, relativist, constructivist and hermeneutic points of view" (p. 29; for references, see the authors' paper).

---

[17] As shown, for example, by the pages proposed in reaction to a Google search for "situated" (conducted on February 26, 2002; confirmed on October 2, 2005, and on May 21, 2006).

10.2.   RITTEL'S ARGUMENTATIVE MODEL

An alternative approach to design that merits special mention is the argumentative model proposed by Rittel (1972/1984, p. 325). In spite of the interest that it has evoked in the domain of cognitive design research, Rittel's model has not been applied by other researchers to analyze designers' activity in empirical design studies. That is why it not discussed any further in this book.

Yet, Rittel remains a reference in the domain of cognitive design research because of, at least, two other important contributions. Together with Webber, he proposed the notion of "wicked" problems (1973/1984) (discussed elsewhere in this book). With Kunz, he was at the start of the design-rationale movement, when the authors formulated the IBIS method (issue-based information systems) for structuring and documenting design rationale (Kunz & Rittel, 1970; see also Conklin & Begeman, 1988).

In an interview with Grant and Protzen, in which they discuss "planning societal problems," Rittel (1972/1984) sketches his argumentative model of the design process. According to Rittel, "the act of designing consists in making up one's mind in favour of, or against, various positions on each issue" (p. 325). One may observe this argumentative structure of the planning process if one looks at it as a network of issues, with pros and cons (cf. IBIS' structure in terms of topics, issues, arguments, positions, model problems, and questions of fact). "As distinguished from problems in the natural sciences, which are definable and separable and may have solutions that are findable, the problems of governmental planning—and especially those of social or policy planning—are ill-defined; and they rely upon elusive political judgment for resolution" (Rittel & Webber, 1973/1984, p. 136). Rittel and Webber insist that there is no definitive "solution" to such problems: "Social problems are never solved. At best they are only re-solved—over and over again" (p. 136).

Rittel (1972/1984) insists that, at least, two areas in design methodology need to be highlighted:

> The further development and refinement of the argumentative model of the design process, and the study of the logic of the reasoning of the designer[, that is,] the rules of asking questions, generating information, and arriving at judgments. . . . The second area of emphasis should be work on practical procedures for implementing the argumentative model: The instrumental versions of the model. (pp. 323–324)

Rittel qualified his argumentative model as a "second-generation design method" relative to the first-generation "systematic design methods" movement. User involvement in design decisions and the identification of users' objectives were the main characteristics of this second-generation movement. In this spirit, which participated in a more general, democratically oriented political current, Cross organized in 1971 the Design Participation Conference (Bayazit, 2004).

Without proposing especially an argumentative model of design, several authors have analyzed argumentative activities and their role in design (e.g., Baker, Détienne, Lund, & Séjourné, in press; Fischer, Lemke, McCall, & Morch, 1991; Martin et al., 2001).

10.3. SITUATIVITY AND COGNITIVE ERGONOMICS

Cognitive ergonomics, in any case as practiced in France, is inherently situated—even if not in the sense of the "classical" situativity approach. It has been thus since its origin and so embedded is this viewpoint that most researchers in the domain until recently did not even feel the need to use this qualification explicitly.

Francophone cognitive ergonomics is situated in that one of its main premises is that the characteristics of the situations in which people are working are quintessential for the understanding and modeling—and modification—of these people's situations. Relying heavily on "work analysis,"[18] it mainly conducts its studies in people's actual workplaces, observing and examining both people while involved in their actual work, and people's work situation. Referring to Leontiev, forerunner of activity theory, Francophone ergonomists have introduced the distinction between "task" and "activity" (see the earlier section *Definitions in the Domain of Data Collection*), which continues to be essential in Francophone ergonomics research (Desnoyers & Daniellou, 1989; see also Marmaras & Pavard, 1999; and the collection of papers in Leplat, 1992b, 1992c).

In the conclusion of their presentation of the SELF, the Francophone Ergonomics Society, Desnoyers and Daniellou (1989) write:

> Although Francophone Ergonomists have insisted on the specificity of their approach to work, they are quite aware and interested in convergent approaches developed elsewhere. . . . Cognitive Anthropology, Ethnographic Description, Situation Awareness, Situated Action are . . . domains Francophone Ergonomists feel close to. . . . De Montmollin [1995] has proposed that we should refer to this type of Ergonomics under the name of Ergonomics of Activity. The new denomination has the advantage of describing the field in a more functional manner.

The special issue of the electronic journal @*ctivités* (2004), "Activité et Action/Cognition Située" [Activity and Situated Action/Cognition], critically discusses the contributions of situativity approaches, and details their "debts" to previous work, especially ethnomethodology (Garfinkel, 1967, quoted in Relieu,

---

[18] "Work analysis can be described as addressing three phases: the description of the actual work situation, that of the activity of the operator, that of the effects of this activity both on the operator and the work system" (Desnoyers & Daniellou, 1989). In ergonomics, an "operator" is a professional whose work is analyzed (cf. the Lugano Convention, which defines an operator as a person who controls the operations of a facility; cf. also Note 3).

Salembier, & Theureau, 2004) and conversational analysis (Sacks, Schegloff, & Jefferson, 1974, quoted in Relieu et al., 2004). In their Introduction, the editors of the special issue also present explicitly the proposals and conclusions that authors from the domain of ergonomics, especially Francophone ergonomists, had already formulated previously to the onset of the situativity movement.

10.4.   SITUATIVITY, DESIGNING, AND PLANNING

There exist some particular relations between situativity, designing, and planning:

- In one of its forms, planning is itself a design activity (see our later subsection *Temporal and Spatial Constraints*).
- A comparison between, on the one hand, designers' planning their activity and, on the other hand, their actual organizing their activity, shows that design has an opportunistic organization (see, our later section *The Opportunistic Organization of Design: Decomposition and Planning*).
- In the book *Plans and Situated Actions* by the anthropologist Suchman (1987), one of the forerunners of the situativity approach, planning plays a fundamental role in the introduction and discussion of situativity-related ideas and notions.

Referring to Suchman's book, authors have often viewed the notion of "situated action" as *opposed to* planning and the use of plans. In her "Response to Vera and Simon," Suchman (1993) declares that she does not agree with such an interpretation of her approach, but states that the relation between plans and "the actions they project" "constitutes a central, unanswered question for existing accounts of practical reasoning and action" (p. 75). "Planning is itself a form of situated action . . . that results in projections that bear some interesting, and as yet unexplicated, relation to the actions they project" (p. 72). In Suchman's view, plans are, however, not the "control structures" that they have often been considered to be—in AI research, we should want to add—, that is, they "are not determining . . . the actions they project, at least not in the strong sense of 'determining'" (p. 74). At the CSCW'96 conference, Suchman (as quoted in Bardram, 1997) "commented that an unfortunate, but typical, misreading of her work was that plans do not exist. Plans do exist and should be viewed as [—Bardram quotes Suchman—] 'an artifact of our *reasoning about* action, not... the generative *mechanism* of action' (p. 39, emphasis in original)" (Bardram, 1997, p. 18).

In a critical discussion of Suchman's position, Bardram (1997) shows how plans, however, "do play an essential role in realising work" (p. 17). Based on experiences with the design of a computer system that supports the collaboration within a hospital, the author discusses "how plans themselves are made out of situated action, and in return are realised in situ" (p. 17). This leads Bardram to propose, "work can be characterised as situated planning" (p. 17). Bardram places this approach to work and planning in the framework of activity theory, which emphasizes, according to the author, "the connection between plans and the contextual conditions for realising these plans in actual work" (p. 17). He writes:

> Suchman (1987) shows the importance of differentiating between work and representations of work like plans and process models. Plans are representations of situated actions produced in the course of action and therefore they become resources for the work rather than they in any strong sense determine its course. Suchman emphasizes action as essential situated and *ad hoc* improvisations, which consequently make plans rational anticipations, before the act, and *post hoc* reconstructions, afterward. The theoretical work on situated action, and the studies underlying it, seems to have attained so much attention that the importance of plans and protocols as guidance of work has been neglected. (p. 18)

Bardram (1997) presents the example of medical work, where "*pre-hoc* representations of work like plans, checklists, schedules, protocols, work programs etc. have proved extremely valuable as mechanisms giving order to work" (p. 18). Comparable observations are made by "Schmidt and Simone (1996) [who] raise the rhetoric question to Suchman of 'What is it that makes plans such as production schedules, office procedures, classification schemes, etc. useful in the first place? What makes them 'resources'?' (p. 169)" (p. 18).

Like Bardram, we have analyzed plans as representations functioning as resources that may guide one's activity. In our research, we have confronted designers' plans with the actual organization of their design activity. We have shown that designers use plans as resources for organizing their activity. These plans are, however, not the only resource that guides their action, and they are indeed not *the* control structure. The actual organization of designers' activity depends not only on plans, but also on other opportunities-providing knowledge sources (for more details, see the later section *The Opportunistic Organization of Design: Decomposition and Planning*).

11 Early SIT-Inspired Research

Except for Rittel, Schön is, as far as we know, the first author after Simon to introduce a new approach to cognitive design theory. Another author of early SIT-inspired research is Bucciarelli, who has focused, in particular, on collaborative design analyzed from a social perspective.

11.1. SCHÖN: DESIGN AS A REFLECTIVE PRACTICE

Schön formulated his view on design in terms of "reflective activity" and related notions, especially "reflective practice," "reflection-in-action," and "knowing-in-action." We interpret the underlying activities as forms of what situativity authors have qualified as "situated action" and "situated cognition."

"Reflective activity" may be defined as the "activity by which [people] take work itself as an object of reflection" (Falzon et al., 1997, quoted in Mollo & Falzon, 2004, p. 532). Schön (1983) writes:

> When a practitioner reflects in and on his practice, the possible objects of his reflection are as varied as the kinds of phenomena before him and the systems of knowing-in-practice which he brings to them. He may reflect on the tacit norms and appreciations which underlie a judgement, or on the strategies and theories implicit in a pattern of behaviour. He may reflect on the feeling for a situation which has led him to adopt a particular course of action, on the way in which he has framed the problem he is trying to solve, or on the role he has constructed for himself within a larger institutional context. (p. 62)

In "reflection-in-action," "doing and thinking are complementary. Doing extends thinking in the tests, moves, and probes of experimental action, and reflection feeds on doing and its results. Each feeds the other, and each sets boundaries for the other" (Schön, 1983, p. 280).

In a presentation of "Donald Alan Schön (1930–1997)" in *The Encyclopedia of Informal Education*, M. K. Smith (2001) writes that, even if Schön was "trained as a philosopher, . . . it was his concern with the development of reflective practice and learning systems within organizations and communities for which he is remembered." In design circles, one generally refers to Schön as the author who, through his proposal of the reflective-practice concept, offered an alternative to the SIP approach defended by Simon in *Sciences of the Artificial* (Simon, 1969/1999b).

Schön's research and thoughts on design thus originate from an educational perspective. Schön was an educator. He was Ford Professor Emeritus on Urban Studies and Education, and Senior Lecturer in the Departments of Urban Studies and Planning, and Architecture, at the Massachusetts Institute of Technology, from the early 1970s until his death in 1997 (Pakman, 2000, p. 5). Schön's enterprise is concerned with the way in which "professionals think in action" as "reflective practitioners" (Schön, 1983), and with "educating" this reflective practitioner (Schön, 1987a, 1987b).

Relative to the contrast between the "reflection-in-action" that underlies reflective practice, and "school knowledge" (1987a), Schön does not see himself "as saying anything really new at all." He is drawing on "a tradition of reform and criticism which begins with Rousseau and goes on to Pestilotsy and Tolstoy and Dewey and then, as we approach more contemporary times, Alfred Schultz and Lev Vygotsky and Kurt Lewin, Piaget, Wittgenstein and David Hawkins today" (1987a). It is Dewey who introduced the concept of "reflective conversation with the situation" that is the locus of reflection-in-action (see the title of Schön's famous paper "Designing as Reflective Conversation With the Materials of a Design Situation," 1992).

According to Schön (1987a), reflection-in-action is the "kind of artistry that good teachers in their everyday work often display," whereas school knowledge refers to a "molecular" idea of knowledge, to "the view that what we know is a product," and that "the more general and the more theoretical the knowledge, the higher it is." From the school-knowledge perspective, "it is the business of kids to get it, and of the teachers to see that they get it."

Reflection-in-action is the reflective form of knowing-in-action. It is Schön's assumption at the start of his famous 1983 book, *The Reflective Practitioner*, that "competent practitioners usually know more than they can say. They exhibit a kind of knowing in practice, most of which is tacit. . . . Indeed, practitioners themselves often reveal a capacity for reflection on their intuitive knowing in the midst of action and sometimes use this capacity to cope with the unique, uncertain, and conflicted situations of practice" (1983, pp. 8–9).

In order to show the nature of knowing-in-action, Schön (1987a) uses the example of what happens "if you are riding a bicycle, and you begin to fall to the left." People who *know* riding a bicycle will do the right thing when *in situ*, but will often give the wrong answer when asked certain questions, in a classroom or anywhere else, outside of a bike-riding situation. An example of such a question out of context, might be: "If you are riding a bicycle, and you begin to fall to the left, then in order not to fall you must turn your wheel to the ___?" This contrast between "[doing] the right thing when *in situ*" and being unable to answer correctly when not, requires an explanation:

> This capacity to do the right thing . . . exhibiting the more that we know in what we do by the way in which we do it, is what we mean by *knowing-in-action*. And this capacity to respond to surprise through improvisation on the spot is what we mean by *reflection-in-action*. When a teacher turns her attention to giving kids reason to listening what they say, then teaching itself becomes a form of reflection-in action, and we think this formulation helps to describe what it is that constitutes teaching.

Even if not taken from a professional situation, this example illustrates the classical, generally applicable difference between "knowing how" and "knowing that" (Ryle, 1949/1973, pp. 28–40 and *passim*).

For Schön, design was one of a series of activities in domains that involve reflective practice: City planning, engineering, management, and law, but also education, psychotherapy, and medicine. Architectural design was the first professional domain studied by Schön in order to develop his epistemology of professional practice based on the concepts of reflection-in-action and knowledge-in-action. In his 1983 book, Schön has "collected a sample of vignettes of practice, concentrating on episodes in which a senior practitioner tries to help a junior one learn to do something. . . . The heart of this study is an analysis of the distinctive structure of reflection-in-action" (pp. 8–9). Indeed, the characteristics of design that Schön presented as general were displayed in the communicative context that he used to collect his observations, that is, educational situations. Focusing on the education of reflective practitioners in the domain of design, Schön's studies examined design students learning with experienced designers (Schön, 1992; Schön & Wiggins, 1992). These studies have been conducted in "reflective practicums such as the design studio in architecture" (Schön, 1987a; see also Schön, 1984).

Adopting ethnographically-inspired or workplace-oriented perspectives (Nilsson, 2005) in his analysis of particular educational design projects, Schön (1983) discusses specific situations in detail, in order to reveal the central role of reflection-in-action in professionals' practice. In their reflective conversations with design situations, designers "frame" and "reframe" problems. In such conversations, "the practitioner's effort to solve the reframed problem yields new discoveries which call for new reflection-in-action. The process spirals through stages of appreciation, action, and reappreciation. The unique and uncertain situation comes to be understood through the attempt to change it" (Schön, 1983). "Furthermore, the practitioners' moves also produce unintended changes which give the situation new meanings. The situation talks back, the practitioner listens, and as he appreciates what he hears, he reframes the situation once again" (Schön, 1983, p. 131–132).

In one of his first papers handling specifically with design (1988), Schön announces that, "in this paper, [he] will treat designing not primarily as a form of 'problem solving,' 'information processing,' or 'search'" (p. 182). Problem solving is generally considered as handling problems as "given," whereas the process of "problem setting" is neglected. Starting with problems as "given," matters of "choice or decision are solved through the selection, from available means, of the one best suited to established ends. But with this emphasis on problem solving, we ignore problem setting, the process by which we define the decision to be made, the ends to be achieved, and the means that may be chosen. In real-world practice, problems do not present themselves to the practitioner as givens. They must be constructed from the materials of problematic situations which are puzzling, troubling, and uncertain" (1983, pp. 39–40). "Problem setting is a process in which, interactively, we *name* the things to which we will attend and *frame* the context in which we will attend to them" (Schön, 1983, p. 40; the emphasis is ours).

Naming, framing, moving, and evaluating are central in Schön's view of design. As we see later, one of the advances of current SIT-inspired research is the operationalization of these and other notions that are central in reflective practice.

For Schön, his observations and his approach to these observations "should be contrasted with the familiar image of designing as 'search within a problem space'. . . . The designer *constructs* the design world within which he/she sets the dimensions of his/her problem space, and invents the moves by which he/she attempts to find solutions" (Schön, 1992, p. 11).

An example of problem setting in architectural design is the following. Problem setting occurs when architects see the project on which they work in a new way: For example, they see a T-form figure as two L-form figures back to back.

Another design characteristic, introduced through an example from architectural design, is the "seeing-moving-seeing" sequence, which is applied iteratively on "design snippets" (Schön & Wiggins, 1992).[19] It consists of action sequences such as observing a drawing, transforming it, and, observing the result, discover "certain unintended consequences" of the transformation move (p. 139). Architects may indeed have a certain intention in transforming a drawing, but they are generally unaware of all possible consequences of their actions. Their intention is liable to evolve in their conversation with the drawing. Referring to Simon, Schön notices that it is because of our "limited awareness" and our "limited ability to manage complexity" that designing has this "conversational structure of seeing-moving-seeing" (Schön & Wiggins, 1992, p. 143). Schön and Wiggins refer several times to *Sciences of the Artificial*, in which Simon introduced his idea of human limited information-processing capacity into the theory of designing. They emphasize, for example, that people, therefore, "cannot, in advance of making a particular move, consider all the consequences and qualities [they] may eventually consider relevant to its evaluation" (p. 143).

Schön thus notices "the remarkable ability of humans to recognize more in the consequences of their moves than they have expected or described ahead of time" (Schön, 1992, p. 7). As pointed out long ago by the urban designer Christopher Alexander, who is also quoted by Schön, "our ability to recognize qualities of a spatial configuration does not depend on our being able to give a symbolic description of the rules on the basis of which we recognize them" (Schön, 1992, p. 137). Analogously, and as noticed by Christopher Alexander as well, even if designers are able to make, tacitly, "qualitative judgments," they are not necessarily able to state, that is, to make explicit, the criteria on which they base them (Schön, 1992, p. 138). This observation once again refers to the knowing-in-action as distinguished from reflection-in-action.

---

[19] Goldschmidt (1991) sharpens this idea, distinguishing seeing-as and seeing-that, in a "dialectics of sketching."

11.2.   BUCCIARELLI:
DESIGN AS A SOCIAL PROCESS

The philosopher Schön studied design in educational contexts of student-professor interaction. The engineer Bucciarelli (who has collaborated with his colleague-philosopher, see Schön, 1992) used "participant observation" as method. "Pretending to be an ethnographer while participating as an engineer" in design projects (1984, p. 185), Bucciarelli examines the "reflective practice in engineering design" (the title of his 1984 paper).

Adopting this "ethnographic perspective on engineering design" (the title of his 1988 paper), Bucciarelli considers that "designing is more than a cognitive process, although design knowledge and designers' heuristics are essential ingredients" (1988). He focuses on the collective nature of design, his main premise being that "designing is a social process"—using a formulation also adopted by Schön (1988, p. 182). "Engineering design is the business of a collective or team [whose] different participants, with different competencies, responsibilities and interest 'see' the object of design differently" (Bucciarelli, 2002, p. 219). Design is "a process of negotiating among disciplines" (Bucciarelli, 1988, p. 160, p. 164) (cf. Rittel's view of design as argumentation). The design that results from such a process is a "social construction" (Bucciarelli, 1988, p. 167). Indeed, "a single object can be understood differently within different 'object-worlds'" (Bucciarelli, 1988, p. 161). As inhabitants of different "object-worlds," designers, while admittedly working on the same object of design, see this object differently (Bucciarelli, 2002, p. 220). The author presents a Tower of Babel vision of the design process, which might lead people to "wonder how it succeeds. . . . Simply put, in many instances design does not succeed" (p. 220) "But most failures of process are hidden from view, kept within the memory of the collective, never revealed to the world, and even within the firm, often easily forgotten. Few participants would deny that designing can be better done; the process improved" (p. 221).

12 Current SIT-Inspired Research

It is mainly on a critical basis with respect to Schön—especially in relation to his lack of precision regarding the operationalization of the notions he proposed—that current SIT-inspired research is being developed (Adams et al., 2003; Dorst, 1997; Dorst & Dijkhuis, 1995; Roozenburg & Dorst, 1999; Valkenburg, 2001; Valkenburg & Dorst, 1998).

Dorst and Dijkhuis compare Simon's and Schön's "fundamentally different paradigms" (Dorst, 1997, p. 204), focusing on Schön's "design-as-experienced" view. They consider that "design is not just a process or a profession," "it is *experienced as a situation* that a designer finds him/herself in" (Dorst & Dijkhuis, 1995, p. 264).

Dorst (1997) was one of the first authors to undertake the Simon-Schön comparison. In an empirical study, he evaluated the ability of each paradigm to describe "integration," which Dorst considered as "one of the key issues of design-as-experienced." "Someone is designing in an integrated manner when he/she displays a reasoning process building up a network of decisions concerning a topic (part of the problem or solution), while taking account of different contexts" (Roozenburg & Dorst, 1999, p. 34).

Dorst (1997) concludes that the two perspectives are differentially accurate and fruitful for describing and understanding different design phases. Examined from Simon's perspective, a designer works as if design involves "objective interpretation." From Schön's viewpoint, a designer looks at design as if it involves "subjective interpretation." Schön's perspective is most appropriate for conceptual design, less so for preceding phases (Roozenburg & Dorst, 1999, p. 35). For detail design, it will probably also "lose some of its descriptive power" (Roozenburg & Dorst, p. 36). For the information phase, Simon's paradigm performs better. An analysis of the individual Delft protocol (Cross et al., 1996) leads Dorst and Dijkhuis (1995) to conclude that it is also appropriate for the embodiment phase.

In an empirical study, Dorst and Dijkhuis (1995) examine how accurately descriptions of design based on Simon's and on Schön's approaches each capture the activity "as experienced by the designers themselves" (p. 261). "Distilling" data-processing systems from both paradigms, the authors use the resulting descriptions to compare the "descriptive value" of each paradigm. Considering it essential that a description method preserves the link between process and content in design decisions, the authors evaluate how each system is able to do so.

The results corresponding to Simon's rational problem-solving system are presented in four separate graphs (corresponding to acts, goals, topics, and contexts). Dorst and Dijkhuis (1995) conclude that, using this system, it is difficult to relate process and content. "Textual analysis (concentrated on the content of the design problem, and consequently outside this paradigm) remains necessary to forge them solidly" (p. 270).

The results of the reflection-in-action system are exemplified in a descriptive summary of the protocol. This allows Dorst and Dijkhuis (1995) to narratively link process and content for three central components of the system, namely moves, frames, and "underlying background theory" "corresponding with the personal view of the designer on design problems and his/her personal goals" (pp. 271–272). The authors summarize the protocol "by only taking into account the 'successful' moves and frames (the ones that the designer stayed with). This gives a clear picture of the what, how and why of the design concept" (p. 272). Dorst and Dijkhuis thus conclude that, in contrast to the rational problem-solving system, the reflection-in-action paradigm makes it possible to relate the design process and its content (p. 274). We discuss this conclusion in a later section.

Valkenburg and Dorst (1998) investigate the "suitability" of the "mechanism of reflective practice" to describe team designing. Using protocol analysis, they compare two design teams with respect to Schön's four central reflective-practice activities in design: Naming (i.e., "naming the relevant factors in the situation"), framing (i.e., "framing a problem in a certain way"), moving (i.e., "making moves toward a solution," i.e., "the actual designing"), and reflecting" (i.e., "evaluating these moves"). The authors describe the general mechanism of reflective practice as used by designers working in a team situation, as an articulation between these four activities.[20] "The designers start by naming the relevant issues in the design situation, framing the problem in a certain way, making moves toward a solution, and reflecting on those moves and the current frame. Reflection is a conscious and rational action that can lead to reframing the problem (when the frame is not satisfactory), the making of new moves, or attending to new issues (naming, when the reflection leads to satisfaction)" (Valkenburg and Dorst, 1998, p. 254).

The two design teams vary in their success. Valkenburg and Dorst attribute the disparity to the differences in "effective design time" spent on the four activities. Compared to their unsuccessful colleagues, the successful team spends less time on naming (3% vs. 49%), more time on moving (73% vs. 39%) and reflecting (21% vs. 8%). However, both teams spend the same small amount of time on framing (only some 3% to 4% of their time). One often refers to Schön because of his emphasis on framing in design, but here this activity thus occupies quantitatively a marginal position. Yet, the successful team consecutively develops five different frames, whereas the unsuccessful team develops only one frame.

In a review of several experimental studies conducted on engineering students, Adams et al. (2003) propose operationalization of the two notions "problem setting" and "engaging in a reflective conversation [with a situation] listening to [the] situation's back-talk" (p. 281, p. 286). Using verbal-protocol data, the authors compare entering and graduating students.

---

[20] It seems to us that this description may also hold for individually conducted design.

Problem setting is captured through three of its attributes. The first two translate "breadth of problem perception" (Adams et al., 2003, p. 285), namely designers listing design factors that are important and gathering information as they engage in solving a design problem. The third one is transitioning between these activities during designing.

Each of the activities proposed as operationalization of problem setting correlates positively with measures of quality, thus providing substance to this notion. The graduating students "display more problem setting behaviors, and therefore are potentially acting more like professionals" (Adams et al., 2003, p. 285). Compared to their freshmen colleagues, they define more broadly the problem they are solving and transition more frequently among design activities:

- They list more design factors and cover a larger portion of the problem-definition space.

- They gather more pieces of information and cover more categories (even if they also neglect important problem elements).

- Throughout the design process, they often return to the problem-setting activities of problem definition and gathering information (pp. 284–285).

Their transitions are "suggestive of a kind of conversation between problem setting and problem solving activities—perhaps suggesting a structure for successful reflection-in-action" (p. 285).

According to Schön as read by Adams et al. (2003), reflection-in-action can be "triggered by surprises that interrupt the flow of skilled, practiced performance, and shift the designer's thinking to a more conscious mode of analysis" (p. 277). It is via the "back-talk" with the surprising situations and encouraged by them, that surprises lead to reflection-in-action. Based on an analysis of the key characteristics of reflection-in-action, the authors qualify design as a particular, goal-directed "iterative process," purposefully progressing "through stages of the design process to revisit and address design issues" (p. 286). "The central cognitive activities triggering an iteration [include] self-monitoring, clarifying, and examining activities" that most likely "result in redefining problem elements and coupling revisions across problem and solution elements" (pp. 286–287).

Adams et al. (2003) qualify the iterations as "transformative": In a "dialectical interaction between representing the problem and specifying the solution," the students reach "new understandings," which they integrate into the continually evolving representation of their design task (p. 288). These transformative processes represent "conceptual shifts" in the students' thinking "and as such may facilitate and mark design learning" (p. 288). In the opinion of the authors, they are "indicative of what Schön refers to as engaging in a situation's back-talk" (p. 288).

Compared to freshmen, seniors were more likely to engage in such transformative processes and activities that trigger shifts in their mode of analysis, thus accomplishing redefinition of problem-solution elements and coupled revisions across these elements.

Again, each of these activities positively correlated with measures of quality and with information-gathering behavior (a problem-setting attribute). As Adams et al. (2003) write:

> In particular, the amount of time in transformative processes significantly correlated with the amount of information gathered (p<0.05) and with the number of transitions (p<0.05). This data suggests that not only are seniors more likely to engage in iterative activities that may be indicative of reflective practice, but that these activities are associated with indirect measures of effective practice. (p. 289)

Based on their results, Adams et al. (2003) conclude, "problem setting and engaging in a reflective conversation across problem setting and problem solving activities are important features of effective design practice" (p. 292). They also observe that there is a dialectical interaction between problem understanding and gathering information, which is "typical of complex and ambiguous design tasks" (p. 292).

Valkenburg (2001), in her analysis of Schön's framework, concludes that "not all designing is done in a reflective practice way." In her empirical studies on team design, she also observes "other ways of working": A "wait-and-see attitude" and "[dealing] with the design problem as a set of separate design issues."

Gero is an author who has been conducting research on design for many years. He has played an important, instigative role in the domain of "artificial intelligence in design" (Gero, 1991, 1992). More recently, Gero has also adopted situativity-related notions, analyzing, for example, "conceptual designing as a sequence of situated acts" (Gero, 1998a). The models he develops include concepts as "emergence and processes which match those of exploration" (Gero, 1998a, p. 168). "Constructive memory" is another important concept for Gero (1999a, 1999b). In order to clarify the underlying notions, he refers to Dewey, to whom he attributes the origin of the idea. He quotes Dewey "via Clancey: 'Sequences of acts are composed such that subsequent experiences categorize and hence give meaning to what was experienced before'" (Gero, 1999a, p. 32).

In spite of its debts to Dewey (1859–1952), this view on memory does not correspond to the ideas advocated by psychology authors in the 1950s and 1960s. Yet, in contemporary cognitive psychology, memory is indeed considered to be constructed, and to evolve continuously, to be a process rather than a fixed state. Nowadays, memory is generally no longer considered as a storehouse—even if not every cognitive psychologist will agree with the description that Gero (2002) gives of "constructive memory," according to which "memory . . . must be newly constructed every time there needs to be a memory" (p. 5).

In this book, we do not discuss any further Gero's work on situated design. It is mainly concerned with general, theoretical analyses, aiming computational models (of design, of creativity in design) and artificial agents (Gero, 1998c; Gero & Kannengiesser, 2004; G. J. Smith & Gero, 2005; Sosa & Gero, 2003).

13 Discussion of the Situativity Approach to Design

We distinguish the early and more recent SIT-inspired research.

Our critique of early SIT-inspired work concerns:

- Its character with respect to theoretical paradigms, which is quite allusive.
- The definition of its central notions, which lacks precision.
- Its methods for data collection, analysis, and modeling, which offer no tools to derive higher-level descriptions from the data (see also Nardi, 1996, p. 83).
- The character of its conclusions, which is anecdotal. Early sit-inspired studies often stop short by merely presenting raw data, rather than providing results likely to be replicable and conclusions likely to be generalizable across situations.

The early SIT-inspired studies present extremely rich descriptions of *unique* designers implementing *unique* activities in *unique* situations.

As remarked by Nardi (1996, pp. 93–94) in her comparison of activity theory, situativity theory, and distributed cognition, SIT models "do not account very well for observed regularities and durable, stable phenomena that span individual situations." This remark pertains completely to Schön's approach to design. "Schön, himself, once described reflective practitioner research as 'non rigorous inquiry' (Schön, 1987, p. 3)," as spotted by McMahon (1988) in a paper discussing the similarities between the theoretical conceptions of reflective practice and action research. Schön indeed did not analyze the structure of "reflective practice" — and neither did Bucciarelli. Both authors proposed detailed, narrative presentations of the design projects they observed — and, for Bucciarelli, in which the author participated — conveying the vivacious flavor of the richness of all activities that such a project may involve.

In his review of *The Reflective Practitioner*, I. Alexander (2001)[21] regrets that Schön, who provides interesting analyses of specific work situations, "which he discusses in detail to bring out how reflection fits into the professionals' use of knowledge," does not "go on to elaborate the structure, rules, and techniques needed to conduct the process efficiently. . . . He disappointingly stops short of doing this." Alexander quotes Schön himself:

> Reflection-in-action is a kind of experimenting. . . . In what sense, if any, is there rigor in on-the-spot experiment? . . . Questions such as these point to a further elaboration of reflection-in-action as an epistemology of practice. One might try to answer them by appeal to a structure of inquiry, but we do not know what such a structure might be or how it might be discovered.

---

[21] Not to be confused with the famous urban-design researcher and theorist Christopher Alexander, also referred to in this text.

Alexander comments: "And [Schön] goes on to say that he'll look for answers in the documented examples."

Schön formulated his ideas, for an important part, as a "critique on the short-comings of Technical rationality" (Roozenburg & Dorst, 1999). Yet, Roozenburg and Dorst (1999) remark that "much of [this critique] had already been raised within the design methodology community a decade earlier, especially by Rittel and Webber (1973/1984) and Hillier, Musgrove and O'Sullivan (1972)" (p. 36). For the SIT-inspired design methodologists Roozenburg and Dorst, "the chief importance of Schön's work is his plea for the emancipation of design in overly knowledge-oriented universities" (p. 29). The authors criticize the lack of precision in Schön's theoretical framework. They consider that, "as an empirical theory, Schön's theory of reflective practice, as it stands, is . . . weak and fuzzy. . . . One looks in vain for explicit definitions of his central concepts" (p. 40). With respect to Dorst's (1997) empirical study, Roozenburg and Dorst notice:

> [It was] difficult to draw general conclusions from the description of design as a reflective practice. The treatment of design as reflection-in-action lacked the clarity and rigor that was achieved by the rational problem-solving paradigm. Because of the weakness of the underlying theory, for example, the problem of finding clear empirical references for Schön's general concepts, it was deemed premature to draw general conclusions from this description of design. (p. 36)

Roozenburg and Dorst conclude, "Schön does not come very far in explicating 'the rigour in its own terms' that he ascribes to the process" (p. 40).

Another, more specific criticism formulated by these contemporary SIT-inspired researchers is that Schön "overstates the role of framing in the context of design" (Roozenburg & Dorst, 1999, p. 39). They add, "criteria for 'good' frames are largely missing" (p. 40). This remark may be related to Valkenburg and Dorst's (1998) observation that the two design teams that they studied spent nearly no time on framing (only some 3% to 4% of their time). Roozenburg and Dorst attribute Schön's overemphasis on framing to his ignoring, what the authors call, "the dual nature of material products," the fact that artifacts are "both socio-cultural and physical constructs. . . . The physical world seems to offer much less room for different subjective knowledge constructs than the socio-cultural world" (p. 39). Valkenburg (2001) also notices "the [difficulty] to unambiguously define what a frame is." She does not manage to "recognize frames, but [is able] to recognize framing." Yet, Valkenburg concludes that "the reflective practice way is, up till now, still the only way . . . to describe the developing design content within projects, while linking this content to the design process and designers."

Schön has been influential in cognitive design research. Nevertheless, Adams et al. (2003) consider that his reflective-practitioner model has not had "a significant or broad influence on engineering design and engineering-design education" (p. 277). Among other evidence, the authors mention that they identified only one reference in a search of journal articles in the INSPEC database, commonly used in engineering (p. 277).

A last critical remark concerning Schön's conceptual framework is that his ideas about design were based on design students interacting with experienced designers, rather than on design professionals working for themselves, or designing together. They concerned "reflection-on-action" rather than reflection-in-action (Valkenburg, 2001).

In SIT-inspired studies that are more recent, Dorst and other researchers have presented analyses and have formulated theoretical frameworks with more precision. Several authors have started to operationalize the central notions in Schön's work. Their studies constitute examples of ways to directly and indirectly capture and measure these notions in empirical data. Adams et al.'s (2003) work substantiates some of Schön's ideas that had been inspiring, but rather vague in that they could cover extremely diverse behavioral phenomena. Research like theirs opens up a way to develop Schön's framework at a theoretical level, and to counter criticisms such as those formulated concerning the early SIT research!

The following remark specifically concerns the design-as-experienced approach as adopted by Dorst and Dijkhuis (1995). Research in cognitive psychology has shown that, for complex cognitive activities, such as those involved in design, the way in which people experience their activity does not necessarily correspond to the way in which they actually proceed. This has been observed, for example, with respect to the organization of designers' activity. We have shown that the way in which designers think that they are going to organize or have organized their activity, does not correspond to its actual organization (Visser, 1994a; Visser & Hoc, 1990).[22] This remark is not meant to suggest that description of people's way of "experiencing" their activity is not interesting or may not be relevant—but that people's experienced activity does not necessarily coincide with their actual activity.

Finally, we want to mention that, even if SIT-inspired researchers consider, by definition, that a particularly relevant component in a design situation is the interactional situation with design colleagues and other people, as far as we know, there have not yet been proposals of specific SIT models of designers' interactional activities with other design participants.

In spite of our criticisms with respect to situativity, we consider essential the contributions of this approach! The insistence on the role of constructive aspects of design, such as designers' not simply modifying ideas based on new data, but also adopting new interpretations of "old" data, concurs with our view on design as the construction of representations. As claimed by Schön (1988): Designing is "a kind of *making*. . . . What designers make . . . are *representations* of things to be built" (p. 182).

---

[22] See also Stacey and Eckert (2003, Note 5, p. 179) who consider that "Dorst and Dijkhuis's (1995) comparison of the views of design thinking put forward by [Simon and Schön] doesn't do adequate justice to the information processing theory position."

IV

Confronting Classical and Modern Views on Design

This part confronts the classical SIP and the modern SIT-inspired views on design. It positions the confrontation in the larger context in which SIP versus SIT debates have been conducted in the cognitive sciences.

Outline of this part. This part first presents two explicitly conducted SIP-versus-SIT debates, focusing on the famous 1993 *Cognitive Science* special issue on situated action (SA). It then closes with a confrontation between the two approaches that focuses on the authors discussed in the two previous parts (Parts II and III).

14 Symbolic Information Processing Versus Situativity Debates in the Literature

Several publications have been the place of a symbolic information processing versus situated cognition/situated action debate, especially the famous 1993 special issue of *Cognitive Science* on situated action (1993) and the discussion in the *Educational Researcher* between, on one side, Anderson, Reder, and Simon (1996, 1997) and, on the other, Greeno (1997). In this book, we concentrate on the first debate. The second one focuses on situated learning and education, a minor theme as regards our present preoccupations. The structure of the *Educational Researcher* debate merits, however, a short presentation.

The *Educational Researcher*'s editor, Donmoyer (1996), notices that Anderson et al. (1996) argue that "advocates of a situated cognition approach to learning and teaching often overstate the empirical support for their position and underestimate the empirical support for rival points of view" (p. 4). According to Anderson et al., "much of what is claimed by [the 'situated learning' movement] is not 'theoretically sound'" (p. 5). In their paper, Anderson et al. review what they consider the central claims of situated learning with respect to education, and cite, in each case, "empirical literature to show that the claims are overstated and that some of the educational implications that have been taken from these claims are misguided" (p. 5).

In his response to this critique, Greeno (1997) first notes that Anderson et al. incorrectly qualify the propositions they contest, as "claims of situated learning." Second, he considers that the empirical evidence advanced by Anderson et al. in order to show the defects of situativity, is, to the contrary, "compatible with the framing assumptions of situativity; therefore, deciding between the perspectives will involve broader considerations than those presented in [Anderson et al.'s] article" (p. 5).

As we see throughout this chapter, this situation is typical of the SIP-SIT debate—and possibly of many, if not most, debates, be they theoretical, political, or still of another nature. The proponents of each position present what they consider "evidence" showing the defects of their opponents' view. In response, these opponents refute this "evidence," or refute its relevance.

In his Introduction to the SIP-SIT discussion in *Cognitive Science*, Norman, 1993) writes:

> This issue ... contains a debate among proponents of two distinct approaches to the study of human cognition. One approach, the tradition upon which cognitive science was founded, is that of symbolic processing, represented in the article by Alonso Vera and Herbert Simon [which triggered the debate]. The other more recent approach, emphasizing the role of the environment, the context, the social and cultural setting, and the situations in which actors find themselves, is variously called situated action or situated cognition. It is represented by four articles written

in response to Vera and Simon: James Greeno and Joyce L. Moore; Philip Agre; Lucy Suchman; and William Clancey. (p. 1)

Among the topics that are central in the debate, we have selected the questions that are most relevant here. "What is at issue?" is the first one. "What is SA?" receives an answer not only by its proponents, but also by its opponents— entertaining or even leading to several disagreements. "What is a symbol?" plays a similar role. Related to these questions, we have identified two couples of critical distinctions: Denotation versus interpretation, and recognition versus direct perception.

14.1.   WHAT IS AT ISSUE?

In their discussion of SA, Vera and Simon (1993b) contest its claims and present symbolic systems that, in their opinion, perform well in the situations that, according to the situativity authors, require SA. This leads Vera and Simon to conclude, "the systems usually regarded as exemplifying SA are thoroughly symbolic (and representational)" (p. 7).

An example is Agre and Chapman's (1987) video game program Pengi, which Vera and Simon, in their opening paper, discuss as one of the "research projects that have been viewed as exemplifying SA principles" (Vera & Simon, 1993b, p. 24). However, for Vera and Simon, Pengi "satisfies the definition of a symbol system" (p. 37): It seems to have "categorical representations of states in the world and functional characterizations of those states" (p. 37).

Agre (1993) considers that "Vera and Simon fail to distinguish between worldviews and theories. A 'worldview' is a largely unarticulated system of vocabulary, methods, and values shared by a research community. Suchman's (1987) worldview is ethnomethodological; Lave's (1988) is dialectical; [Agre's] might be called interactionist. A 'theory' is some substantive proposition formulated within a worldview. You argue against a theory by presenting contrary evidence. But a given worldview will admit a wide variety of alternative theories" (p. 62). For Agre (1993), Pengi has arisen "from a dissatisfaction with STRIPS-like planners and with world models" (p. 67). In line with his analysis of Vera and Simon's positions as a "worldview" rather than a theory, he replies to Vera and Simon in the following terms: "Whether Pengi's hardware can be reconstructed in 'symbolic' terms is beside the point; the point is that we interpret it in interactionist terms, and we hold that these are the best terms for analyzing the phenomena" (p. 68). For Agre (1993), "the critical issue is whether one's categories locate things in agents and worlds separately [as does the 'symbolic worldview' represented by Vera and Simon] or in the relationships between them" as do the SA authors—even if they all "provide different accounts of what the latter might mean" (p. 69).

For Greeno and Moore (1993),

[the question] seems to be something like this: Whether (1) to treat cognition that involves symbols as a special case of cognitive activity, or (2) to treat situated activity as a special case of cognitive activity, with the assumption that symbolic processing is fundamental in all cognitive activity. [They] advocate the first option; Vera and Simon advocate the second. (p. 50)

For Clancey (1993),

symbolic models have explanatory value as psychological descriptions. . . . But arguing that all behavior can be *represented* as symbolic models [as do Vera and Simon] misses the point: We can model everything symbolically. However, what is the residue? What can people do that computer programs cannot? What remains to be replicated? (p. 89)

14.2.   WHAT IS "SITUATIVITY"?

The different authors participating in the debate adopt different definitions of this central notion.

For Vera and Simon, the two authors defending the SIP approach, a sensible use of the term "situated action" refers to action that occurs "in the face of severe real-time requirements" and that is "based on rather meager representations of the situation" (Vera & Simon, 1993b, p. 41). Vera and Simon judge that all action requires "some internal representation of the situation—perhaps minimal in the case of situated action, more elaborate in the case of planned behavior when fewer unexpected events occur" (p. 41).

As noted previously, Vera and Simon (1993b) consider that they can give a symbolic interpretation of SA. "Some past and present symbolic systems are SA systems. The symbolic systems appropriate to tasks calling for situated action do, however, have special characteristics that are interesting in their own right" (p. 8). "The term 'situated action' can best serve as a name for those symbolic systems that are specifically designated to operate adaptively in real time in complex environments" (p. 47). Vera and Simon even judge that "the goals set forth by the proponents of SA can be attained only within the framework of symbolic systems" (p. 7).

Vera and Simon's opponents in the special issue clearly do not agree with them. For Greeno and Moore (1993), who have proposed the term "situativity," the corresponding theory is a large movement that refers to "emerging scientific practices, empirical findings, and theory [that] include the development of ecological psychology, the ethnographic study of activity, and philosophical situation theory" (p. 49).

The opinion attributed by Vera and Simon (1993b, pp. 7–8) to SA proponents, according to which SA researchers deny that "symbolic processing lies at the heart of intelligence" is indeed endorsed by Greeno and Moore (1993, p. 50).

However, as noted by these SA authors, "the issue hinges crucially on the meaning and theoretical status of the concept of *symbol*" (p. 49). They contend:

> The central claim of situativity theory is that cognitive activities should be understood primarily as interactions between agents and physical systems and with other people. Symbols are often important parts of the situations that people interact with [but] we expect that accounts of most—perhaps all—individual and social cognitive phenomena will include hypotheses about processes that are not symbolic. (pp. 49–50)

14.3.   WHAT IS A "SYMBOL"?

For Vera and Simon (1993b), "symbols are patterns," which means that "pairs of them can be compared" (p. 9). Their physical nature is "irrelevant to their role in behavior" (p. 9). They can be "of any kind: Numerical, verbal, visual or auditory" (Simon, 2001a, p. 205). "Symbolic," thus, "is not synonymous with 'verbal'" (Vera & Simon, 1993b, p. 10). Crucial is that "we call patterns symbols when they can designate or denote" (Vera & Simon, 1993b, p. 9).

In order to act in an environment, people use perceptual and motor processes that connect "symbol systems" with their environment. With respect to these symbol systems, Vera & Simon (1993b) write that they are built from "symbols, which may be formed into symbol structures by means of a set of relations" (p. 8). In addition,

> a symbol system has a memory capable of storing and retaining symbols and symbol structures, and has a set of information processes that form symbol structures as a function of sensory stimuli, which produce symbol structures that cause motor actions and modify symbol structures in memory in a variety of ways. (p. 9)

The perceptual and motor processes that connect a symbol system with its environment provide the system "with its semantics, the operational definitions of its symbols. Evocation of a symbol by stimuli emanating from the thing or situation it designates also provides the system with access to (some or all of) the information stored in memory about the thing designated. The memory is an indexed encyclopedia" (Vera & Simon, 1993b, pp. 9–10).

For Greeno and Moore (1993), "a symbol or symbolic expression is a structure—physical or mental—that is interpreted as a representation of something" (p. 50). The authors consider that the distinction between symbolic and nonsymbolic processes depends on "whether a process includes a semantic interpretation of a symbolic expression, that is an interpretation that gives the symbolic expression referential meaning" (p. 54).

According to Greeno and Moore (1993), Vera and Simon's definition of symbol is inappropriate, whereas their own use of the term is "consistent with a long tradition in philosophy, psychology, and linguistics. Examples include Dewey's (1938) distinction between *signs* and *symbols*, and Peirce's (1902/1955) distinction between *indices* and *symbols*" (p. 50).

For Vera and Simon (1993a), however, Greeno and Moore's is "an alternative, albeit rather restrictive definition of symbols, but it is not the one that cognitive scientists working within the symbolic-system hypothesis have used" (p. 78). In their short "Reply to reviewers," Vera and Simon do not name any of these cognitive scientists. An example of a French researcher is Le Ny (1989a), who in his research on verbal understanding indeed adopts the same stand as Vera and Simon. For this view of symbols, Le Ny refers to formal semantics, and to Peirce interpreted in a carnapian and tarskian vision.

Agre considers Vera and Simon's definitions of symbols and symbol systems as "metaphors." "There exist many reasonable uses of 'symbol,' each embedded within a worldview" (p. 64).

14.4. DENOTATION VERSUS INTERPRETATION, AND RECOGNITION VERSUS DIRECT PERCEPTION

"Situativity" and "symbols" are both viewed in different ways by the authors from the two sides. "Symbols," however, seem to be the main element of controversy (see, e.g., Greeno and Moore's remark).

As we saw previously, for Vera and Simon (1993b), symbols "designate or denote" (p. 9). Greeno and Moore (1993) take an additional step and claim, "symbolic expressions" are structures that are "interpreted as [representations] of something" (p. 50).

The distinction between "denotation" (or "designation") and "interpretation" seems essential in this debate. It is related to another crucial distinction, the one between "recognition" and "direct perception"—itself related to the conscious and/or verbalizable character of processing.

According to Newell and Simon (1972),

> a symbol structure *designates* (equivalently, *references* or *points to*) an object if there exists information processes that admit the symbol structure as input and either (a) affect the object; or (b) produce, as output, symbol structures that depend on the object. . . . The relation between a designating symbol and the object it points to can have any degree of directness or indirectness. (p. 21)

For Vera and Simon (1993b), on the one hand, "symbolic theories generally make no specific assumptions about what part of the processing takes place at a conscious level and what part is unconscious, except that symbols held in short-term memory (in the focus of attention) are generally available to consciousness, and often can be reported verbally" (p. 10). On the other hand,

> awareness has nothing to do with whether something is represented symbolically, or in some other way, or not at all. It has to do with whether or not particular symbols are available to consciousness in short-term memory. Thus, in an act of rec-

ognition, the symbol denoting the object recognized is consciously available, the symbols denoting the features that led to the recognition generally are not. The recognizer is aware of the former but not of the latter. (p. 19)

"Only a small part of denotation entails direct perception of objects and their behavior, as the denotation of 'cat' mainly does. The connections between a symbol structure and its denotation can be complex and highly indirect (e.g., the denotations of concepts like 'empirically true,' or even 'China')" (Vera & Simon, 1993b, p. 12). In the same way, as Vera and Simon (1993c, p. 128) notice in their reply to Clancey (1993), the "rules" used in a computer system that simulates human activity are not necessarily accessible to, nor verbalizable by, the person whose activity is being simulated.

Thus, in Vera and Simon's view of "symbolic processing," many cognitive activities are not conscious, and *yet* proceed by underlying symbolic processes. In Greeno and Moore's (1993) view of "symbolic processing," many cognitive activities are not conscious, *and consequently do not proceed* by underlying symbolic processes.

Notice that this opposition is not merely terminological. Greeno and Moore (1993) use the terms "direct perception" and "recognition" as distinctive notions (for whose meaning they refer to Gibson, 1966, and Neisser, 1989, 1992). They consider "recognition" a symbolic process, as do Vera and Simon, but "direct perception" is no symbolic process for Greeno and Moore (in Gibson's view, it is indeed unmediated by mental activity). Vera and Simon use "recognition" as a technical term with a precise definition. We do not know, however, if their use of "direct perception" also conveys such a specific, technical sense. Whatever may be, in their "Reply to Reviewers," Vera and Simon (1993a) affirm that "the problem with creating a distinction between direct perception and recognition is that there seems to be no principled way of deciding what sorts of phenomena fall into which category" (p. 79).

So, when Greeno and Moore (1993) state, "there is more than 'symbolic processing'," one can only agree—providing that one adopts the authors' definition of "symbols." Indeed, not all processing requires conscious, semantic, interpretational activities.

We suspect that a related misunderstanding concerns the association between "information" and meaningful, verbalizable content.

Simon (1992a) considers that "the currently popular ideas of situated action and situated learning" have the following "simple essence." "In those frequent cases in everyday experience in which attention is directed primarily to the external environment, the cues in this constantly changing scene, which are constantly modified by the actions of the system, provide the triggers for evoking the successive productions" (p. 132). This evocation proceeds through "recognition, in which cues permit the recognition of familiar situations or objects and thereby gain access to the information about them that is stored in memory" (p. 132). He continues:

A principal reason why [many unrigorous procedures are] satisfactory is that people are in constant interaction with their environment. We take a step, notice new conditions and constraints that our reasoning ignored, and adjust our next step accordingly. In designing, we record our tentative design decisions in a drawing, and the drawing reveals interactions that were ignored in our premises. . . . Here again, is situated action arising from the operation of a production system. (p. 134)

Reader's of this book may remember that we consider that the SIP framework attributes too important a role to recognition!

15 Confrontation of the SIP and SIT Approaches

In his Introduction to the *Cognitive Science* special issue, Norman (1993) states that, in his view, the "two traditions do not seem to be contradictory.... They do not conflict.... They emphasize different behaviors and different methods of study" (p. 3). However,

> the social structure of science is such that individual scientists will justify the claims for a new approach by emphasizing the flaws of the old, as well as the virtues and goodness of the new. Similarly, other scientists will justify the continuation of the traditional method by minimizing its current difficulties and by discounting the powers or even the novelty of the new. (p. 3)

Vera and Simon (1993b), the two authors representing the SIP approach, argue, for example, that "the new approach could easily be incorporated within the old" (Norman, 1993, p. 1). For Anderson et al. (1996), Vera and Simon's (1993b) review is "in support of the mutual compatibility of modern information processing theory and situated cognition" (p. 5).

In a paper from 1978, Simon made a statement concerning the scope of the SIP approach that sounded like a more modest claim. He stated "information-processing theories have made especially good progress in providing explanations for solving relatively well-structured, puzzle-like problems of the sorts that have been most commonly studied in the psychological laboratory" (p. 272).

Situativity-oriented authors judge that the *conditions* of sense giving merit more attention and more study than the possibly *underlying activities*—be they symbol-based or not. They focus on the social and interactional aspects of these conditions. For Anderson et al. (1997), "the situated movement has performed a valuable service in emphasizing the important contextual and social aspects of cognition. However, the situated position has not shown that it provides the right theoretical or experimental tools for understanding social cognition" (p. 20). This remark may be related to our critique of the early SIT-inspired research. Yet, the *Cognitive Science* debate dates from more than 10 years ago, when there was not yet modern SIT-inspired research.

Two of the proponents of the SIT approach, Greeno and Moore (1993), agree with their "opponents" Vera and Simon "that 'breaking completely' from symbolic cognitive theories would be the wrong thing to do, but [they] believe that something like 'departing fundamentally' is required" (p. 57). "Within the historical development of psychology," the authors perceive, "in the present situation, a prospect of completing a dialectical cycle, in which stimulus-response theory was a thesis, symbolic information-processing theory was its antithesis, and situativity theory will be their synthesis" (p. 57).

In their reply to Clancey (1993), Vera and Simon (1993c) note:

Complex systems are best described in several levels, each level corresponding to a different time frame. In the case of human behavior, events in the range from a few milliseconds to a few tens of milliseconds are best described in a vocabulary of neurons and their actions [by neuropsychologists]; events in the range from a few hundreds of milliseconds to a few minutes are best described in the operation of a symbol system with a large memory, relatively limited input and output devices, and a modest capability of learning (within this time frame) [by information-processing psychologists]. Events in the range from hours upward are best described in terms of collections of interacting symbol systems that influence not only each others' behaviors but also each others' memory stores [by social psychologists or sociologists]. (p. 130)

SIP and SIT researchers are perhaps at different levels!

Confronting SIP and SIT is not evident. In their comparative analysis of Schön's design-as-experienced and Simon's rational problem-solving views, Dorst and Dijkhuis (1995) concluded that the rational problem-solving system was not appropriate to relate process and content in design decisions. Using different types of description formalisms, the authors' comparison was like one between apples and oranges. In addition, we consider that the different types of description formats used by the authors (separately presented graphs vs. narratively connected descriptions) may have induced their conclusion. It seems difficult to separate the influence of the paradigms from the influence of the description formats.

According to Dorst and Dijkhuis (1995), an analysis focusing on design problems' content is "outside" the rational problem-solving paradigm (p. 270). Yet, in our understanding, none of the SIP characteristics predisposes this framework to a separation between process and content (see also Ericsson & Simon, 1980, 1984/1993). By the way, working in an SIP-compatible approach, the EIFFEL team has elaborated a method that uses a *predicate(arguments)* structure for encoding design dialogues, thus establishing systematically a link between process *(predicate)* and content *(arguments)*. This is the COMET method, developed for the analysis of collaborative design projects based on designers' verbal interactions (Darses et al., 2001).[23]

Founded on his empirically based comparison between Simon's and Schön's paradigms, Dorst (1997) has observed that the two paradigms differ in their appropriateness to describe different phases of design. This observation has led him to conclude that the choice of a research or modeling paradigm depends on one's research goals and objects of study, and, most important, on the kind of design activity one aims to study.

As we have shown in our presentation and discussion of the SIT theoretical framework, we consider early and current SIT approaches as rather different. Besides, we noticed that current SIT-inspired research is being developed by

---

[23] In a recent analysis of data collected in a collective architectural design meeting (Détienne & Traverso, 2003), we establish quite explicitly this link in the form of an analysis that shows designers' moving forward in their design proposals, through an articulation between solutions and the activities that introduce and modify them (Visser, in press).

their authors out of a critical position with respect to Schön, and that they indeed have furthered the SIT position. As far as we know, there is no fundamentally new approach inside the SIP framework. Yet, we consider that essential cognitive aspects of design are neglected by both the SIP framework and the SIT approach (in both its early and current form).

Simon, in his discussion of design, disregards the rich and specific characteristics of the activity pointed out previously. Schön leaves us with detailed descriptions of extremely "rich" situations, but does not make the reflective step (N.B.) that we consider critical, that is, required in order to get to a more general level that yields statements about design in general.

We agree with authors such as Schön concerning the importance of problem setting, framing, and reframing, concerning the active and constructive aspects of problem understanding, which lead to a continuously evolving problem representation. We concur with Schön, who considers as obvious that "problem structuring" and "problem solving" are intertwined — something that Simon's SIP approach discards. We also agree, however, with SIP-oriented authors such as Simon in their insistence on the importance of people's internal representations, which, even if they evolve under the influence of people's interaction with their environment (human and artificial, social and material), provide people with a "basis" from which they may act in and on this environment. Situativity researchers do not deny the existence of such an internal apparatus, but they do not posit it as a basis and they do not give it much attention — this also holds for current SIT-inspired researchers, who do not seem to consider particularly significant "purely" cognitive factors.

Each research community chooses its focus of interest for research — at the cost of simplifying and ignoring other topics. Studies conducted by reference to the SIP paradigm pay more attention to people's *use of knowledge and representations* in problem solving than to the *construction of representations*, and they do not analyze activities as they occur in interaction with other people and the broader environment. SIT-inspired authors focus on the *consequences* of people's interaction and the influence of the environment, the social and cultural setting, and the situations in which people find themselves, but they usually neglect the underlying cognitive structures and activities.

Interested as they are in different aspects of designing, SIP and SIT researchers necessarily are also silent concerning different aspects! SIT researchers state that designers' "dominant resource" is the situational context. For SIP researchers, this resource seems to be the range of designers' problem-solving methods. Even if the SIP approach does not deny that these resources may evolve under the influence of designers' interaction with their environment, it does not pay much attention to these factors, which play a central role in SIT-related approaches. If SIT approaches do not deny the existence of internal resources, it is only recently that they have begun to analyze them. We saw that the SIT paradigm — especially, in its reflective-practice form — is evolving towards theoretical frameworks that are more substantial.

One might be tempted to conclude with Norman (1993) that the "two traditions do not seem to be contradictory," but that "they emphasize different behaviors and different methods of study" (p. 3). Indeed, for a designer confronted with a design task, resources for action—the elaboration of representations, strategies, and other action—depend on both internal (knowledge and representations) and external (social and artifactual) data. We judge, however, that these two types of data are not in a symmetric relation: It is the *designer* who, *using her or his knowledge and representational activities*, establishes the relationship between internal and external data. It is because these links are to be established by way of representation that we attribute such an essential role to the *construction of representations—by the designer*!

V

Design as Construction
of Representations

The three previous parts have introduced and compared two families of approaches to design, namely the classical SIP, and the alternative SIT-inspired frameworks. Part V presents our own approach to design. It has two main grounds, its theoretical and inspirational background, and its empirical basis.

From a theoretical viewpoint, we both propose new elements and adopt elements from the aforementioned two frameworks. We introduce a quality that is not exclusive to SIP, but yet is particularly characteristic of this classical approach to cognition. That is the precise and explicitly proceeding approach to research. This was not typical of SIT-inspired approaches, but current SIT-inspired research is also aiming in this direction.

A specific SIP view that we endorse is to consider problems as particular task representations, that is, to analyze people's tasks in terms of the representations they construct of these tasks. We also adopt a position that Simon introduced, and to which SIT-inspired researchers agree as well. This is the viewpoint of design as satisficing. With respect to the view of design as a type of cognitive activity rather than a professional status, we suppose that it is not rejected by SIT researchers, but is not considered relevant either.

From the SIT approach, we retain especially the emphasis on the active and constructive aspects of designing and on the evolving characteristics of representations. As noted in the section *Situativity and Cognitive Ergonomics*, the general emphasis on the importance of the situation in which a person is performing is not specific to the situativity approach. It is a particularly central position in Francophone ergonomics. So is the attention to both people's task and their activity. SIT-inspired researchers emphasize what Francophone ergonomists would qualify as an activity-oriented view on design, but they do not theorize the distinction between task and activity.

In our own research, from the beginning and continuing today, an important accent has been on the task-activity opposition and the relevant data-collection methods that go with it. We therefore, especially, adopted real-time observation in designers' professional context (field studies), focusing on the role of the various information sources and the further situational factors in their activity. It is also mainly because of our focus on designers' activity that our approach is strongly grounded in the analysis of data from empirical design studies, performed mainly in professional work settings.

This way of data collection might have made our research resemble studies conducted from a situativity standpoint. However, as it has always been guided by a theoretical framework, our selection and analysis of data did not concur with the spirit of situativity. On the other hand, having started with an SIP perspective to design (analyzing design as problem solving), our work has deviated more and more from this approach, as this book shows.

Specific to our position is the central role attributed to representation, both the representation activity and the internal and external structures that are being constructed.

Outline of this part. We start by a comparison between definitions of design proposed in the literature, and comment on the central elements in our own definition. The next chapters present representational structures (from representations at the source to representations at the end of a design project), followed by a chapter that introduces representational activities. This separation between structures that are constructed and the corresponding construction activities is of course a purely analytical one. In the concluding *Discussion* chapter of this part, we propose some links between the two. We further defend the assumption that there are different "forms" of design, even if we also adhere to the "generic design" view, that is, design is a specific cognitive activity, distinct from other types of cognitive activity.

16 Definition of Design

Definitions are representations: They focus on *aspects* of the object they aim to cover—even if their authors imagine that their focus is the object's *essence*. In our following review of definitions, we restrict ourselves to cognitive aspects of design.

Even considered from such a perspective, the characteristics of design that are selected as essential may still differ. Our focus on the activity of design further orients our view. Definitions may thus focus on characteristics whose relevance we do not deny, but that do not inform us about cognitive aspects of designing. An example is the definition by Moran and Carroll (1996b): "The primary goal of design is to give shape to an artifact—the product of design. The artifact is the result of a complex of activities—the design process" (p. 1).

Many definitions of design focus on the result of the activity, that is, the artifact product, ignoring the nature of the activity. In their *wording*, they may use references to actions, such as "specifying," "defining," or "creating," but not detail any activity in developments of the definition. Another characterization by Moran and Carroll (1996b, p. 13) considers design as "the process of creating tangible artifacts to meet intangible human needs" (p. 2), to which the authors add, "creating and constructing" are "the defining acts of design." There are authors, such as Stacey and Eckert (2003, p. 164), who view designing as "modeling." Both are positions close to ours, but they present no further specification of the cognitive aspects of the activity. Other authors, often from AI-related communities, consider design as a constraint-satisfaction activity, but propose methods without any cognitive underpinnings (see Darses, 1990a, for a cognitive-psychology discussion of this approach).

Designers are not to produce the artifact product, but its specifications. We consider essential to distinguish between these specifications and the artifact product itself. A group of definitions seems to neglect this difference. They qualify design, for example, as "the creation of artifacts that are used to achieve some goal" (Mayall, 1979, in his *Principles in Design*, referred to in Atwood et al., 2002).

For authors focusing on the specifications, design consists of producing plans or descriptions, or still other forms of representations of the artifact product (Archer, 1965/1984; D. Brown & Chandrasekaran, 1989; De Vries, 1994; Hoc, 1988a; Jeffries et al., 1981; Kitchenham & Carn, 1990; Schön, 1988; Whitefield, 1989). Applied to software design, for example, this means that design leads to a plan that allows transformation of these specifications into executable code (Jeffries et al., 1981; Kitchenham & Carn, 1990). Many empirical studies of "software *design*" focus, however, on elaboration of executable code—that is, coding—rather than design.

According to most definitions, the artifact product has to meet certain requirements, that is, accomplish certain functions, fulfill certain needs, satisfy certain constraints, allow attaining certain objectives, and possess certain characteristics. Designing is thus usually defined—even if implicitly—as a goal-oriented activity—even if this goal is not fixed, or preestablished.

After a presentation of our definition of design, this chapter presents a characteristic of design that we consider essential, that is, its creative nature.

16.1. OUR DEFINITION OF DESIGN

Globally characterized, from our viewpoint, design consists in specifying an artifact (the artifact product), *given requirements* that indicate—generally neither explicitly, nor completely—one or more functions to be fulfilled, and needs and goals to be satisfied by the artifact, under certain conditions (expressed by constraints). At a cognitive level, this specification activity consists of constructing (generating, transforming, and evaluating) representations of the artifact *until they are so precise, concrete, and detailed* that the resulting representations—the "specifications"—specify explicitly and completely the implementation of the artifact product. This construction is iterative: Many intermediate representations are generated, transformed, and evaluated, prior to delivery of the specifications that constitute the final design representation of the artifact product together with its implementation. The difference between the final and the intermediate artifacts (representations) is a question of degree of specification, completeness, and abstraction (concretization and precision). A similar view is expressed by Goel (1995), who writes: "Design, at some very abstract level, is the process of transforming one set of representations (the design brief) into another set of representations (the contract documents)" (p. 128).

Our focus on the activity and the intermediate representational structures should not lead to forgetting the central role of both the requirements as source and the implementable specifications as goal, that together steer both activity and representations. There are other activities that construct representations (especially, the interpretation of semiotic expressions), but due to their having other types of inputs and outputs than design, the underlying activities differ as well (cf. B. Hayes-Roth, Hayes-Roth, Rosenschein, & Cammarata's, 1979, distinction between generation and interpretation problems, discussed later).

In our core definition, we qualify design as *construction*, rather than *transformation* of representations, because

- "transformation" may convey the connotation of the representations to be transformed, being *given* (cf. The misunderstanding with respect to problems being given discussed earlier).
- "construction" is more general: We see as we proceed that it involves both generation and transformation activities (and it also requires evaluation).

16.2.   CREATIVITY

The creative character of design is one of its essential characteristics, which has, however, not been highlighted by many authors in the domain of cognitive psychology and cognitive ergonomics (except for Archer, 1965/1984; Cross, 1997, 2001a; Van der Lugt, 2000, 2002; see also refs in Cross, 2001a). Simon (1999a), in his Foreword to Dym and Little's *Engineering Design* (1999), qualifies design and creation as equivalent activities.

Archer (1965/1984) considers that "design involves . . . the presence of a creative step" (p. 58). "Arriving at a solution by strict calculation is not regarded as designing" (p. 58). "Some sense of originality is also essential" (p. 58). However, both characteristics have a relative character:

> Just how much originality is needed to distinguish the preparation of working details from actual designing, and just how little is inevitably required to distinguish calculation from designing, is very difficult to prescribe, especially in the field of mechanical, electronic, and structural engineering. The satisfaction of these two conditions together is sometimes referred to as the presence of a creative step. (pp. 58–59; see also De Vries, 1994)

Even if we consider creativity an essential characteristic of design, there is a difference between design and creation, in the sense of artistic creation. There is a difference between "an architect preparing plans for a house" or "a typographer preparing a layout for a page of print"—both clearly designing—and "a sculptor shaping a figure." The sculptor is generally involved in an artistic creative activity, that is, creating, not designing (see Archer, 1965/1984, p. 58). Two necessary elements for qualifying an activity as "design" rather than "creation" are the "prior formulation of a prescription or model" and "the hope or expectation of ultimate embodiment as an artifact" (Archer, 1965/1984, p. 58) (cf. our emphasis on the central role of both the design requirements and the artifact's specifications). A third difference that we want to add is that the specifications formulated by a designer should be so explicit (precise, concrete, and detailed) that other people than the designer be able to implement them. An artist may produce a sketch, a cartoon, but, on the one hand, this plan does not need to be precise or complete and, on the other hand, the final sculpture or text may be even completely different than the one specified!

Mayer (1999) analyzes the definitions of "creativity" proposed by the different authors who contribute to the *Handbook of Creativity* edited by Sternberg (1999). His conclusion is that "the majority [of authors] endorse the idea that creativity involves the creation of an original and useful product" (p. 449) (cf. Barron's, 1955, view that creativity in ideas reflects "novelty that is useful," quoted in Van der Lugt, 2003). Mayer emphasizes that "there is lack of consensus on such basic clarifying issues as whether creativity refers to a product, a process, or person; whether creativity is personal or social; whether creativity is common or rare; whether creativity is domain-general or domain-specific; and whether creativity is quantitative or qualitative" (p. 451).

In our characterization of design as "creative," we use the notion to qualify the activity rather than its result. We use the terms "novelty" and "innovative" to qualify a result that is particularly new and original. Novelty may be the outcome of an activity that is not creative.

Cross (2001a) analyzes the "creative leap" and the "sudden mental insight" (notion for which he quotes Ö. Akin and Akin, 1996), which he considers both as essential in creative thinking. He proposes the notion of "creative event" as the equivalent of the sudden mental insight, for the abrupt, unexpected evocation of a new concept that immediately becomes the design participants' focus. Based on an analysis of the way in which people introduce such a new concept, Cross (2001a) proposes to analyze this creative event, not as a creative "leap," but as a "bridge" between problem space and solution space (see also Cross, 1997). "The crucial factor [in a creative event], the 'creative leap,' is the bridging of [the two partial models of the problem and solution] by the articulation of a concept . . . which enables the models to be mapped onto each other. The 'creative leap' is not so much a leap across the chasm between analysis and synthesis, as the throwing of a bridge across the chasm between problem and solution. The 'bridge' recognizably embodies satisfactory relationships between problem and solution" (Cross, 1997, quoted in Cross, 2001a, pp. 89–90). Cross (2001a) concludes that, rather than regarding creative thinking as mysterious, we now possess "better descriptors of what actually happens in creative design" (p. 97). He mentions problem framing (such as described by Schön, among others), coevolution of problem and solution, and conceptual bridging between problem space and solution space.

A final remark concerns the intentional character of creativity. Creative thinking may appear a deliberate activity that one is happy to engage in. In his study of engineering-design projects, Marples (1961) concluded, however, that the difficulties of the problems to be solved *compelled* designers to "innovate." The elaboration upon the initial solution proposals required hard and ingenious thinking. In the projects analyzed by Marples, "innovation [was] imposed on the [designer] and not sought by him" (p. 67).

17 Representations

"Knowledge" and "representation" are two central concepts in cognition. There have therefore been many debates around "representation" in the cognitive sciences, for example, in relation to "representation versus no representation" (Clancey, 1991; "Dreyfus and Representationalism," 2002; Greco, 1995a, 1995b; Paton & Neilson, 1999; Peterson, 1996; "Representation," 1995; Robinson & Bannon, 1991). Coming from completely different origins, R. A. Brooks (1991) and Dreyfus (2002) have been claiming, each one in a paper titled "Intelligence Without Representation," that intelligent behavior does not require representation.[24] In this book, we clearly claim that intelligent behavior *does* require representation, and that people construct and use representations. Another discussion concerns the "symbolic versus nonsymbolic representation." We come back to this issue, based on Goel's (1995) discussion of "sketches."

Some authors use the term "representation" only in the case of a well-structured formulation (see F. Heylighen, 1988, and other references in that text). That does clearly not correspond to our use of the term.

We wish to stress that, due to its prefix re-, the term "representation" itself is somewhat misleading—the German and Dutch terms "Vorstellung" and "voorstelling" (in English, "presentation") are less deceptive. "Representation," as we view it, should not be interpreted literally: There is no "original," no independent "reality" beyond one's experience to which the representation corresponds. Contrary to representations in AI, the representations constructed by human systems are not more or less faithful replicas of an entity that is re-presented, be it internally or externally.

This position is close to Von Glasersfeld's (1981) "radical constructivism." The two main principles of constructivism are: "(1) knowledge is not passively received but actively built up by the cognizing subject; (2) the function of cognition is adaptive and serves the organization of the experiential world, not the discovery of ontological reality" (Von Glasersfeld, 1989; cf. also Piaget).

17.1. FORMS OF REPRESENTATIONS

Many distinctions are possible here. In the context of designing, particularly relevant are those between internal (mental) and external; initial, intermediate, and final; visual-spatial and verbal-sequential; and private and jointly used representations (cf. what we qualify as "interdesigner compatible representations").

We allot special sections to the discussion of several of these representational forms. There is, for example, a section concerning external representations, because of their centrality in design research, but this special consideration

---

[24] Vera and Simon (1993a) in their "Reply to Reviewers" in the 1993 special issue of *Cognitive Science* on situated action assert that R. A. Brooks' robots were actually using representations.

should not lead to neglect the importance of internal representations (see, e.g., research on software schemata, Détienne, 2002a). Even if mental representations have received much attention in traditional cognitive-psychology research, we do not possess much data on their possibly specific characteristics in design. It is not because people use external representations that they do not use internal representations. In their "reflective conversations with design situations" (Schön, 1992), for example, designers use internal representations—even if Schön does not highlight this aspect. Using external representations requires internal representations, in a continuous interaction between the two types of representations. As both are being constructed using knowledge, this also emphasizes the central role of knowledge in design.

Representations based on spoken, verbal expressions. Relative to other representations, those based on oral, generally verbal, expressions have at least two specific, noteworthy characteristics.

First, speaking does not allow drafts: Once pronounced, speech is definitive (cf. gestures).

Second, even if one may, of course, record speech, people in normal interactional situations do have no other trace of their interlocutors' statements than their memory (see also J. C. Tang & Leifer, 1988). By contrast, an interesting characteristic of external representations is that they generally leave traces—even if many exploratory drawings are thrown away (McGown et al., 1998). Later on, designers can come back to these residual representations, examine them at ease, and show them to colleagues (cf. research in the domain of design rationale concerning the way and the form in which it may be useful to keep traces of design; see Buckingham Shum & Hammond, 1994; Moran & Carroll, 1996a).

Nevertheless, oral interaction plays an important role in design. Oral verbal interaction, articulated or not with graphical and gestural representations, plays a central role in co-design meetings (Darses et al., 2001; D'Astous et al., 1998; Détienne, Visser, D'Astous, & Robillard, 1999).

17.2.   FUNCTIONS OF REPRESENTATIONS

Representations have many different functions and may be used for various aims. Working by means of representations is an interactive process in which the representations one constructs both depend on, and influence the view of one's task. This view evolves, also because of the representations used—whose evolution (through transformation) goes together with one's evolving view. Briefly characterized, representations are tools—cognitive artifacts—

- to keep track of ideas, inferences made, and results and conclusions achieved—be they partial and intermediate, or complete and final.

- to advance understanding and interpretation, and possibly "see" things differently.
- to reach "new" ideas based on "new" interpretations of the representations.
- to derive implications from results already obtained and presented in the representations.
- "to think about situations that [one is] not in or that may not yet exist," "to reason about situations that [one is] unable to experience directly," and to "rely on a kind of 'hypothetical reality' that anchors [one's] reasoning" (Greeno & Hall, 1997).
- to organize one's "continuing work" (Greeno & Hall, 1997; Do, Gross, Neiman, & Zimring, 2000, also mention this organizational function).
- to communicate one's results or conclusions—be they final or intermediate—to other people.

We detail these different functions.

17.3. SYMBOLIC REPRESENTATIONS

Regarding the symbolic nature of representations, we refer to Goel's analysis in *Sketches of Thought* (1995). The author is not "against representation," not even "against symbolic representation." What Goel defends, however, is that the "classical" cognitive-science symbolic representations are not all there is! Especially in ill-structured activities[25] such as design, other types of representation also play an essential role. Based on Goodman's (1969, quoted in Goel, 1995) analysis of symbol systems, Goel explains the essential and typical role played in design by a representational activity qualified as "sketching." Goel's use of this term is not restricted to a particular form of drawing, not even to pictorial symbol systems. In the broad sense that "sketching" has in Goel's work, the term refers to the use of any non-notational symbol system characterized by a certain number of properties. In informal terms, these properties may be qualified as imprecise, fluid, amorphous, indeterminate, and ambiguous (see Table 8.1 in Goel, 1995, p. 182, for the corresponding formal properties, especially their semantically and syntactically dense and ambiguous character).

According to Goel (1995), classical cognitive science is based on what he calls the "Computational Theory of Mind" (CTM), which claims that cognitive processes are computational and require a representational medium (such as Fodor's "Language of Thought") with "some very stringent properties" (p. 3). Among these "CTM properties" are syntactic properties (disjointness, finite differentiation) and semantic properties (unambiguity, disjointness, finite differentiation) (chaps. 7 and 8 *passim*).

---

[25] Goel follows Simon's terminology.

Goel (1995) does not deny that people's performance in certain tasks (e.g., the classical SIP problems) can be successfully explained by reference to mental representations with CTM properties (chap. 3 *passim*, and pp. 75–76). These properties indeed work for well-structured puzzle problems—they do not, however, for ill-structured, open-ended, real-world problems, such as design, planning, scientific discovery—and neither for the arts. Certain symbol systems used for ill-structured problems do not satisfy several CTM properties, especially syntactic disjointness and unambiguity. Mainly in early phases of the activity, such problems also require other symbol systems. Progressing from early to later design phases, designers use each time less non-CTM systems (such as sketching and particular forms of natural language) and each time more CTM-satisfying representational systems (such as drafting and other forms of natural language). In his conclusion, Goel conjectures that these observations do not hold only for designers' external representational symbol systems, but also for their internal, mental systems.

Recently, Goel has started to establish a link between these results and neurological data concerning anatomical dissociations. He establishes, for example, a relation between the processing of ill- and well-structured information and the different cerebral hemispheres.

17.4.   THE OPERATIVE AND GOAL-ORIENTED CHARACTER OF REPRESENTATIONS

Representations are neither "complete" nor "objective." Adapting somewhat Ochanine's (1978) notion of "operative images," we qualify representations as "operative" in that they are, not so much functionally distorted ("distorted" relative to which reference?) and restricted to task-relevant characteristics (Ochanine's view), but *shaped* by these characteristics. Variations in cognitive commitment play an important role in this molding.

Representations are also goal-oriented. They are "constructions, which for some purposes, under certain conditions, used by certain people, in certain situations, may be found useful, not true or false" (Bannon, 1995, p. 67). In more sociopolitically oriented terms, representations are "interpretations in the service of particular interests and purposes" (Suchman, 1995, p. 63). "Even the most seemingly unmediated, veridical representational forms like video recordings do not wear their meanings on their sleeves to be read definitively once and for all" (Suchman, 1995, p. 63). Representational devices are built in order to make things "visible so that they can be seen, talked about, and potentially, manipulated" (Suchman, 1995, p. 63). This visibility is not only for others: An important role in design is played by designers making visible things for themselves, so that, in Schön's terms, they can engage a "conversation" with them, and so advance their design activity.

For Greeno and Hall (1997), representations, "in addition to being representations of something, . . . are for something." The authors examine and practice a view in which they educate pupils to work with representations, especially mathematics, not "just" in order to learn mathematics, but in order to *use* them *for something*, for example, in one of their studies, design. "This contrasts with a common practice in school, wherein students learn to construct representations of information without having a real purpose" (Greeno & Hall, 1997). The reason for which someone wishes to use a representation (e.g., show a decline in a function, or a correlation between two variables) determines, at least in part, the representational form that the person selects. The form depends on the solution process as well. This view also contrasts with a common educational practice, where, in order to solve a problem, students generally are instructed to use standard representational forms and processes depending on the type of the problem.

Greeno and Hall show that people not only use the forms of representation that they have been taught, but also construct novel forms (see research by Hall and other researchers presented in Greeno & Hall, 1997).

Standard notational forms are, of course, also useful, for example, to communicate one's ideas. This quality is due to the existence of widely shared conventions of interpretation. Greeno and Hall (1997) emphasize that teaching the use of such standards to students is thus valuable, but they underline that standard forms have a limitation when their notations are treated as fixed, rather than potential representations that depend on interpretation.

17.5. VIEWPOINTS AND PRIVATE REPRESENTATIONS

In recent years, the notion "viewpoint" has been used in many publications, without having received a clear definition that differentiates it as a particular type of representation. The term is often adopted to refer to the mental representations of design participants that are distinctive due to these participants' specific professional knowledge, domain of expertise, and know-how. However, this acceptance of "viewpoint" does not justify, in our view, the introduction of a specific notion. The representations to which such authors refer do not have characteristics that differ from those generally attributed to "representations." As we have claimed previously, there are no "objective" representations: The influence of people's experience and knowledge, be it professional or otherwise, is one of the factors that influence the nature of their representations. We therefore do not attribute any technical meaning to the term "viewpoint."

We use the term "private representations," when, in the context of collaborative design, we want to distinguish them from the representations jointly used by designers working together—yet, without any theoretical pretension as referring to a particular type of representation. We use the term "personal representation" when the private character of a person's representation is not relevant or established.

17.6.   AMBIGUITY IN REPRESENTATIONS

In their paper "Against Ambiguity," Stacey and Eckert (2003) discuss the ambiguity of representations in design, in the context of what they qualify as the "myth of beneficial ambiguity" (p. 153). As the authors note, the term "ambiguous" is itself ambiguous: "Its standard meaning is 'interpretable in two or more distinct ways'" (p. 153), but it is also used to mean "imprecise." The authors propose to use two different terms for these two meanings: To keep "ambiguity" for the standard acceptation, and use "imprecision" or "uncertainty" to refer to the second meaning.

Representations are not ambiguous or imprecise per se. According to Stacey and Eckert (2003), both ambiguity and imprecision characterize relationships between the representations specified by a designer and the possible artifact products specified by these representations (p. 165). We consider that the relative character of a representation's ambiguity especially depends on the beholder: A representation is ambiguous for the person who perceives at least two interpretations, not for whom only one interpretation exists.

"The myth of beneficial ambiguity" concerns design *communication*. "The widespread belief that ambiguity is beneficial in design communication stems from conceptual confusion" (Stacey & Eckert, 2003, p. 153). This belief is related to two "doctrines." "The first is that ambiguity facilitates creativity by enabling reinterpretation" (p. 154). This may indeed be the case in "early creative design, [in domains such as] architecture, where designers are relatively free of constraints" (p. 155). In more tightly constrained design situations, however, Stacey and Eckert claim that such beneficial effects may not apply. Van der Lugt's research on idea generation in creative problem-solving meetings shows that the relationships between reinterpretation and creativity are indeed rather complicated.

The other doctrine is that "design is inherently social" (Stacey & Eckert, 2003, p. 155). Authors defending the "myth of beneficial ambiguity" in this context generally conflate ambiguity with uncertainty (p. 156). Designers obviously "need to communicate skeletal or incomplete designs" or designs in which some elements or decisions are still provisional. In face-to-face, oral communication, "phrasing and intonation can convey degrees of commitment . . . but visual communication has no such subtle signals built in" (Stacey & Eckert, 2003, p. 167). Yet, in drawings, especially in sketches, roughness may translate the provisional character of designers' mental representations, and may convey this provisionality to designers' colleagues who are to interpret the drawings.

Stacey and Eckert (2003) argue that *if* "ambiguity" has benefits in communication, it is "due to the value (and necessity) of communicating provisional, qualitative, and imprecise designs (as clearly as possible)" (p. 175). It is, however, not the ambiguity as such, but the expression of provisionality that may be valuable.

Therefore, depending on the circumstances, "ambiguity" can be beneficial or harmful. It can lead to the "discovery of useful alternative ideas." However, the authors "suspect that in practice this is rare in design conversations, and very much rarer than worthless misunderstandings" (Stacey & Eckert, 2003, p. 175):

> Ambiguity can be beneficial when the gain from actively clarifying shared understanding is greater than the cost of exploring unacceptable paths. . . . [Whether] interactively refining quick, rough and ambiguous expressions [is] more cost-effective than investing effort in initial clarity . . . depends on the speed and ease with which misunderstandings are corrected and boundaries explored. (p. 176)

The authors' motivation for their battle against ambiguity is Eckert's study of knitwear design. Eckert showed that the communication between designers and technicians constituted "a major bottleneck in the design process." It was "often only partially successful, leading to both inefficiency and inferior products" (Stacey & Eckert, 2003, p. 157).

Stacey and Eckert, in their discussion of ambiguity of representations, focus on intermediate representations as they are used in design communication. Reitman (1964), working in the SIP tradition of the 1960s, adopts another perspective. He takes ambiguity in its standard sense; namely, a problem is ambiguous if it can be interpreted in several ways. He considers that problem ambiguity constitutes a continuum that goes from well-defined formal problems to such ill-defined empirical problems as composing a fugue (the design activity analyzed by the author in Reitman, 1965, quoted in Reitman, 1964).

In Reitman's (1964) view, the ambiguity of an ill-defined problem translates the scarcity or even absence of agreement over a specified community of problem solvers, regarding the referents of the problem's attributes, permissible operations, and their consequences. Even in a particular community, solutions to ill-defined problems are not correct or incorrect, but "accepted" (or rejected)— and only more or less.

The source of this ambiguity is, according to Reitman (1964), the problem's open constraints. An "open constraint" is a constraint "whose definition includes one or more parameters the values of which are left unspecified as the problem is given to the problem-solving system from outside or transmitted within the system over time" (pp. 292–293). The open constraints of ill-defined problems are "surely the most important in characterizing" these problems.

Therefore, the ill-definedness of a task is such with respect to a specified community—and so is its ambiguity.

In a note, Reitman (1964) adds a precision that was suggested to him by Simon. It is not "content area"[26] (e.g., fugue writing or cooking) that determines a problem's ill-definedness. Although the *typical* problem in an area may be either

---

[26] This notion of "content area," used in the context of laboratory problem solving, corresponds to the "domain" in professional situations.

ill- or well-defined, specification of content area by itself does not imply ill- or well-definedness. The example suggested by Simon to illustrate this is the problem "compose the fugue on the score in my desk." Even if, of the basis of its content area, that is, musical composition, one might have thought that it was ill-defined, this problem has only one correct solution: It does not have any open constraints.

Even if there are indeed no open constraints on the problem's *goal state*, its initial state and the constraints on the *permissible operations* are open—something that Simon and Reitman seem to neglect. We suspect that it is why composers may nevertheless be unable to solve the problem—not because they are unable to find *the* solution!

17.7.   PERCEPTS, MENTAL IMAGES, MENTAL MODELS, AND OTHER INTERNAL REPRESENTATIONS

According to Zhang and Norman (1994), different types of representations differentially activate perceptual and cognitive processes. They claim that external representations activate perceptual processes, whereas internal representations usually activate cognitive processes (Zhang & Norman, 1994, p. 118). With Scaife and Rogers (1996), we suppose that things are less systematic, and more complex.

A particular type of internal representations is "percepts," that is, the mental representations that result from perception. They clearly play an important role in design, especially in design of physical artifacts.

In spite of commonalities between imagery (leading to mental images) and perception (leading to—mental—percepts) (see research reviewed in Kavakli & Gero, 2001), there are critical differences between the two processes and their outcomes. Based on a series of experiments, Chambers and Reisberg (1985) conclude:

> One important source of . . . differences between [mental] images and percepts lies in the way each of these comes into being. Perception, initiated by stimulation from an external object, is largely concerned with the interpretation of that object. . . . [Mental] images, in contrast, are constructed as an image of some particular thing or scene. . . . [Mental] images are symbolic. . . . [Mental] images are not *picturelike* [in] that there is no such thing as an ambiguous image. (p. 318)

The experiments presented by the authors show indeed clearly that images are not ambiguous. These results are interpreted by their authors as "arguing against the claim that imagery and perception share a common processing path" (Chambers & Reisberg, p. 326).

Verstijnen and colleagues, in research on mental imagery (Verstijnen, Heylighen, Wagemans, & Neuckermans, 2001; Verstijnen, Van Leeuwen, Goldschmidt, Hamel, & Hennessey, 1998) conclude, "mental images are not inspect-able in the same ways as pictures" (Verstijnen et al., 1998, p. 532). One cannot perform the same operations on such internal images as on external images, and these different images will be differentially useful in creativity (see later discussion).

Various cognitive design studies refer to "mental models" and to other, comparable mental representations that are supposed to preserve, in analogue form, features of the represented entity (see Gentner & Stevens, 1983; Johnson-Laird, 1983; Rumelhart, 1989).

17.8.   INTERDESIGNER COMPATIBLE REPRESENTATIONS IN COLLABORATIVE DESIGN

Given that they incorporate components from various domains of specialty, design projects generally require multiple skills—and collaboration between them.

The role of representations in collective design varies according to its phases. During distributed design, designers each have their own tasks and specific goals to pursue. When co-designing, they have a common goal that they aim to reach by applying their specific skills and expertise. It is then essential that designers, who each also have their personal, possibly private representations, establish what has been qualified by different authors as a "common ground" (Clark & Brennan, 1991) or a "common frame of reference" (De Terssac & Chabaud, 1990; Hoc & Carlier, 2002). Various other notions have been proposed as related, conveying more or less important differences in view: "Shared context," "mutual referential," "mutual awareness," and "mutual manifestness."

These representations concern agreements, especially on the definition of tasks, states of the design, references of central notions, and weights of criteria and constraints. They are often qualified as "common," or "shared," but given the fact that there is no objective reference, we prefer to characterize them as interdesigner compatible representations (Visser, 2006).

The notion of "compatibility" in this context is based on Von Glasersfeld 's constructivist ideas. In a paper on the legacy left by Piaget, Von Glasersfeld (1997) writes the following:

> In order to live in a society, a sufficient number of our ideas—our concepts and schemes of action—have to be compatible with those of others. And this compatibility confers on them a viability that goes beyond the merely individual. The same goes for the acquisition and use of language. Communication with others requires that the meanings we attribute to words prove compatible with those of other speakers.
> Compatibility, however, does not entail the kind of "match" that is implied when people speak of "shared ideas" or "shared knowledge." Compatibility . . . means no

more and no less than to fit within constraints. Consequently, it seems to me that one of the most demanding tasks of AI would be the plausible simulation of an organism's experience of social constraints.

17.9.   EXTERNAL REPRESENTATIONS

In recent years, many cognitive design studies have come to focus on external representations.

In domains where the object of design is a physical artifact, for example, mechanical design, "visual representations are omnipresent throughout the design process, from early sketches to CAD-rendered general arrangement drawings" (McGown et al., 1998). Numerous studies examine their use in mechanical design, but also in architecture and industrial design (Do et al., 2000; Goel, 1995; Kavakli et al., 1998; McGown et al., 1998; Neiman, Gross, & Do, 1999; Purcell, 1998a, 1998b; Rodgers, Green, & McGown, 2000; Scrivener, 1997; Tseng, Scrivener, & Ball, 2002; Verstijnen et al., 1998; Verstijnen et al., 2001).

The important role of external representations is, however, not restricted to domains where the object of design is a physical artifact. In today's software engineering, visual languages and visualization as a development and support tool play an ever more important role. The *Journal of Visual Languages & Computing* and the IEEE Symposium on Visual Languages attest to this fact (see the Visual Language Research Bibliography, devised by M. Burnett and M. Baker, retrieved August 10, 2005, from http://web.engr.oregonstate.edu /~burnett/vpl.html). Some authors in this domain who work on user-interface design, cognitive, and other human-oriented issues, are Blackwell, Burnett, and T. R. G. Green.

Forms of External Representations

External representations may take numerous forms: Visual-spatial, graphical or verbal-sequential; two- or three-dimensional; notes, flowcharts, drawings, plans, or scale models (Nakakoji, Yamamoto, Takada, & Reeves, 2000; Newman & Landay, 2000). Among these different forms, drawings—and especially sketches[27]—have received special attention in cognitive-design studies. These representations play indeed an important role in design, particularly in its early phases. Drawing allows to "think through the end of the pencil" (Purcell, 1998a). In a series of studies on mechanical design, Ullman and colleagues have shown the importance of drawing in this domain of design (Staufer & Ullman, 1988; Ullman, 1992; Ullman & Culley, 1994; Ullman et al., 1987; Ullman et al., 1988; Ullman, Wood, & Craig, 1990). According to analyses reported by Hwang and Ullman (1990), "67% of [the marks made on paper] were drawings or sketches.

---

[27] In the rest of the text, "sketch" is used in its common, restricted sense of a particular type of drawing, and not in Goel's (1995) technical, broad acceptation.

The remaining 33% were text, dimensions and calculations" (p. 343; see also McGown et al., 1998).

"Technical" forms of representation (such as tables, graphs, and equations) are often contrasted with "free" or "artistic" forms of representation (in fields such as painting, sculpture, and literature) and especially with forms that, in addition, allow fluidity and imprecision, such as sketches. Such "creative" forms of representation would be especially flexible; they would be particularly adapted for expressive and communicative use, and open to multiple interpretations (but see Stacey & Eckert, 2003; Van der Lugt, 2000, 2002). Greeno and Hall (1997), however, argue that "representations in mathematics and science also have these properties" and that students might benefit from activities in which they learn to construct and interpret in a flexible way representations that possibly are nonstandard. Such an approach might enable them "to understand and appreciate that mathematical and scientific representations, like those in other domains, are adapted for particular uses" that they perhaps did not expect.

Functions of External Representations

Important functions of external representations depend on the externalization and visualization they allow of preconceptual "ideas" (or "ideas" that will never become concepts), and that may facilitate designing. This holds especially for "graphical," "diagrammatic" (Glasgow, Narayanan, & Chandrasekaran, 1995), or "visual" representations (these different terms are often used for comparable representations). External, graphical representations enable operations on the entity represented that are more difficult or even impossible to perform on internal representations. Using such a representation, for example, it is often easier to "manipulate" an entity, to reason, test hypotheses, and apply other operations on the entity. External, graphical representations often facilitate the discovery or exploration of alternatives, and the prediction of outcomes or consequences of new ideas (Do et al., 2000).

Through the possibilities of simulation, such external representations may be useful in evaluation and in further development of solutions. They serve evaluation in at least two ways. Usually, they are relatively easy to use (compared to internal representations) in order to try out quickly and cheaply different options: The functioning or use of a conceived artifact product may be assessed through its manipulation—even if this operation takes place indirectly, that is, mentally. Juxtaposing various drawings or mock-ups, a designer may compare different possibilities (Do et al., 2000).

These instrumental functions of external representations are essential for designers to advance their design, and work on it through controlled reasoning activities, but also via unintentional and unforeseen discovery. In Schön's terms, the intermediate results of designers' activity, often in the form of external representations, may lead them, in a "reflective conversation with the situation," to evolve in their interpretations, intentions, and ideas for solutions.

External representations—not only graphical—are of course also helpful as memory aids in an extension of internal memory (i.e., in order to temporarily stock provisional ideas, and permanently archive intermediate and final solutions). The storage function of representations concerns not only final representations, nor the straightforward, technical representation of the artifact product, nor the representations communicated to colleagues and other design stakeholders. "As memory for context evaporates over time, supplying sufficient information to enable interpretation is also important for communicating with oneself in the future" (Stacey & Eckert, 2003, p. 163).

Notice that Verstijnen et al. (1998), on the basis of differential results concerning designers' need for sketching, defend that *sketching* is not primarily motivated by memory restrictions (p. 530). Sketching and sketches have other functions. Verstijnen et al. thus contradict the introspective reports of "many artists and designers" who, when asked for their motivation to sketch, "ascribe a function of memory extension to [this] behavior" (p. 530).

Yet, "computational offloading" (i.e., discharging internal working memory) is a function often referred to, even without using this appellation: It refers to the observation that, compared to the use of internal representations, the use of external representations reduces the amount of cognitive effort required to solve informationally equivalent problems (Larkin & Simon, 1987; Zhang, 1997). Indeed, Larkin & Simon have proposed that the perceptual processes that can be applied on external representations allow people to exploit these representations with less effort than the corresponding internal representations, because the grouping together of relevant information in an external representation may make easier processes such as search and recognition.

Graphico-Gestural Representations

In pragmatic linguistics, analysis of graphico-gestural interaction has already a considerable history, but from a cognitive-activity viewpoint, this type of research is just at its beginning (Détienne et al., in press; Détienne & Visser, 2006; Visser, in press; Visser & Détienne, 2005).

In early research analyzing small-group conceptual design sessions, J. C. Tang and Leifer (1988; see also J. C. Tang, 1991) identified the role of gestural activity in "workspace activity." The authors have proposed a framework for the analysis of this activity that establishes relationships between actions that occur in the workspace, and their functions. The "conventional view" of workspace activity considers this space as "primarily a medium for storing information and conveying ideas through listing text and drawing graphics." The authors extend this view, proposing the function of "mediating interaction." In addition to the actions of "drawing" and "listing," they advance "gesture," that is, "purposeful body movements which communicate information, such as referring to existing objects in the workspace or enacting simulations" (p. 245). Besides drawing, which was known to often occur in collaborative-design meetings (46% of the workspace activity, in the authors' analysis), gesture was found to occur frequently: It constituted 35% of the workspace activity, whereas listing made up 19%. The main function of gesture was to mediate interaction between the different design participants: More than half of the gestures (57.5%) served this function through participants engaging their attention.

On the Web site page that presents the research on gesture in her STAR team (Space, Time, and Action Research, retrieved August 16, 2005, from http://www-psych.stanford.edu/~bt/gesture/), B. Tversky notes:

> Although it is typically thought that gestures accompany speech, gestures often accompany listening ... and non-communicative thinking. ... In both cases, they seem to serve to augment spatial working memory, much as sketching a diagram would. ... In collaboration with diagrams, dyads save speech by pointing and tracing on the diagram. Partners look at the diagrams and their hands, not at each other. ... Having a shared diagram to gesture on facilitates establishing common ground and finding a solution. It also augments solution accuracy.

Sketches

In its commonsense, dictionary acceptation, "drawings" are representations of forms or objects on a surface by means of marks; "sketches" are preliminary drawings for later elaboration.[28] Generally, a sketch is drawn by hand; few computerized systems allow a "real" sketching activity producing "real" sketches (but see Decortis, Leclercq, Boulanger, & Safin, 2004; Gross, 1996; Hwang & Ullman, 1990; Leclercq, 1999).

The characteristics considered useful in initial representations, especially their fluidity and imprecision, are typical of this particular form of drawings. Their meaning may be vague and actually change over time (cf. also Stacey & Eckert, 2003, p. 172, concerning sketches' ambiguity).

Nevertheless, sketching is not the only possibility for designers to offer a wide interpretation space for their representations: Stacey and Eckert (2003) affirm that one can also use precise representations such as photographs of artifact products that are similar to the artifact one is aiming (p. 172).

These relatively unstructured, fluid, and imprecise forms of drawings that sketches are, may give access to knowledge not yet retrieved and may evoke new ways of seeing (because of their non-notational properties, according to Goel, 1995). Unforeseen views on the design project in progress are supposed to open up unanticipated potentialities for new aspects or even completely new directions in the design project.

Based on Stacey and Eckert's (2003) critical analysis of the "myth of beneficial ambiguity," one should be conscious of the crucial differences in potential benefit, both between ambiguous representations in early and in later stages of design, and between their use in individual and in collective design.

Sketches' characteristics related to fluidity and imprecision are supposed to result in enhancement of creativity and innovation in design. However, more than imagery (using mental images) or the use of external representations, what

---

[28] We are referring here to "idea-sketches." "Presentation-sketches" are not preliminary generally.

matters for the creative process is that the representations and representational activities allow one to "transform information from one code to another so as to gain multiple perspectives on the task at hand" (Reed, 1993). This is indeed a possibility of drawings, but it is neither an unconditional, nor an exclusive characteristic.

In research on visual, graphic brainstorming, Van der Lugt (2000) has analyzed the claim that sketches result in creativity enhancement. His conclusion that "in early idea generation, sentential, or partly sentential, variations of the brainstorming tool . . . perform stronger than [their] graphic variations" (p. 521) is another argument for the claim that external, graphical representations are not the panacea for creativity enhancement!

Based on these studies (Van der Lugt, 2000) and on research that Van der Lugt (2002) has performed on another visual, graphic brainstorming technique, namely brainsketching, we conjecture that sketching alone is not enough in order to trigger people to be creative: People may, moreover, need to be *stimulated* to engage in *reinterpretation* of their drawings (such as ensured by the brainsketching technique applied in Van der Lugt, 2002; cf. also Verstijnen et al.'s, 1998, results concerning the conditional utility of external representations).

In spite of their fluid character, sketches are not purely idiosyncratic creations. Their meaning is, at least in part, "defined by notational conventions and mediated by the recognition of abstract category memberships, mapping categories of mark-combinations to categories of objects or concepts" (Stacey & Eckert, 2003, p. 171). Designers use notations based on the domain's standard drawing conventions in order to express abstract attributes of a design. They of course personalize their representations by the inclusion of idiosyncratic extensions and variations. This personal selection of graphical symbols seems to be a recurring characteristic of designers' drawing process (McFadzean et al., 1999, referred to in Stacey & Eckert, 2003, p. 171).

18 Representations at the Source of a Design Project: Requirements and "Design Problems"

The representation at the source of a design project expresses the requirements for the artifact product—or, in problem-solving terms, the problem specifications.

In industry or other professional situations, a design project generally starts with client requirements (and input from other company departments, such as product planning, methods, or marketing). As explained in the *Introduction*, qualifying design as a "problem" (the "design problem") refers to designers' representation of their task that consists in satisfying these requirements. It does not convey any presupposition with respect neither to the requirements being exhaustively and unambiguously developed when the designers receive them (Carroll et al., 1997, 1998), nor to a definitive representation of these requirements being constructed by the designers when they first analyze them.

According to the NSF solicitation for its program "Science of Design" concerning software-intensive systems, but that seems applicable also to other domains, "the requirements . . . are rarely complete and the true potential of software-intensive systems is often not realized or understood in the first release of a system. Moreover, these systems are difficult to maintain and adapt to additional requirements."

The "design as a problem" approach has been the object of many SIP-inspired cognitive design research literature (see references in Bayazit, 2004; Cross, 2001a, 2004b; Détienne, 2002a; Eastman, 2001). We therefore discuss in this chapter only the three characteristics on which we propose a complementary or different view. These are design problems' ill-definedness, their complex character, and the way in which the artifact is constrained by design-problem representations.

Before detailing these three characteristics, we discuss design in relation to several problem typologies that characterize design problems relative to other types of problems.

18.1. DESIGN PROBLEMS IN PROBLEM TYPOLOGIES

Cognitive psychology classically distinguishes design problems from transformation problems and structure-induction problems. Two other problem distinctions that are orthogonal to this classical typology are based on the partitioning into "adversary" and "nonadversary," and "semantically rich" and "semantically impoverished" problems. Two related classifications that are adopted in AI are also discussed, namely the "interpretation"-versus-"generation," and the "analysis"-versus-"synthesis" problems typology.

The Classical Cognitive-Psychology Distinction: Transformation, Structure-Induction, and Arrangement Problems

Cognitive psychology distinguishes different types of problems in terms of initial, intermediate, and goal states, and operators. Problems differ depending on the type of representation that the person who is confronted with a task, constructs of the task. In this way, the cognitive-psychology distinction that nowadays is considered the "classical" problem typology, distinguishes transformation (or state-transformation), structure-induction, and arrangement problems. It dates back to the 1970s (Greeno, 1978). Somewhat more recently, design problems have often been presented as constituting the third category—instead of, or combined with arrangement problems (Greeno & Simon, 1988).

Some authors have extended the classical typology. Mayer (1989) proposes two other types of problems: Deduction problems and divergent problems. For Greeno and Simon (1988), deduction is, however, not a problem-solving but a reasoning task.

Greeno (1978) presents his three types of problems as ideal types, and notices that most interesting problems include components of different types. "Design" and "invention" problems are examples of such interesting, mixed types of problems: They may be analyzed as problems that require "inducing structure in arrangement problems" (pp. 263–264). For Greeno, "the most demanding intellectual problems" are "composition problems," which involve "[creation of] an arrangement of ideas whose structure incorporates some significant new understanding" (p. 264). Several studies of painters and Reitman's (1965) famous research on musical composition are referred to as indicating that "most of the intellectual effort involved in composition concerns defining and developing of problems, rather than solving them" (Greeno, 1978, p. 264) (cf. our discussion concerning problem "solving" vs. other problem-related activities). From our point of view, these characteristics apply not only to composition, but to design problems in general, where "defining and developing of problems" may be analyzed as representational activities.

In cognitive psychology, transformation problems are the types of problems that have received by far the most attention. This is mainly due to Simon and colleagues who, working from the SIP viewpoint, examined these problems in the form of play problems (such as the Tower of Hanoi and other classical laboratory tasks), not in the form of tasks performed in professional settings. This focus on transformation problems in the form of play problems has been maintained by other SIP-inspired researchers (in France, see, e.g., Richard, Poitrenaud, & Tijus, 1993).

Discussions by the aforementioned authors of the classical problem classification (and its minor variations) are not particularly instructive about the specifics of design problems. According to Greeno and Simon (1988), in design problems, it is mainly the goal state that is constrained (but not explicitly specified), whereas in transformation problems, initial and goal states are explicitly given and there are constraints on the actions that may be used to achieve the goal state. We elaborate on these differences in our section on design problems' ill-definedness.

One should distinguish a problem category (e.g., structure induction) from a typical example (e.g., diagnosis). For design problems, there seems, for the moment, to be no risk of confusion, given that every specimen of the class of design problems is a design problem. In our Discussion chapter of this part, we open a reflection on the existence of different "forms" of design. Continuing these considerations might lead to the identification of different examples of subclasses of design problems.

Routine Versus Nonroutine Problems

The dimension "routine"–"nonroutine" is commonly used in design studies, but more in the domain of AI than of cognitive psychology. Researchers working in the domain of AI and design (Gero, 1991, 1992, 1998a, 1998c; Logan & Smithers, 1993) have often made this distinction.

Related terms encountered in this literature are "creative," "insightful," "innovative," and "open-ended" design, without the underlying distinctions always being clear (D. Brown & Chandrasekaran, 1989; Logan & Smithers, 1993; Navinchandra, 1991).

In *Exploration and Innovation in Design: Towards a Computational Model*, Navinchandra (1991) opposes "routine" and "nonroutine" design.

D. Brown and Chandrasekaran (1989) focus on the construction of a theory of routine mechanical design. They establish a distinction between "routine" design and "open-ended," "creative" design. They suspect that very little design is in this second class, which is characterized by "extremely innovative behavior" "leading to major inventions or completely new products" (p. 33).

Gero (1990) distinguishes two forms of nonroutine design, namely innovative and creative design.

In between routine and nonroutine design problems, some authors distinguish "variant" design problems, that is, revisions of an existing design (Dym & Little, 1999).

Summarizing the definitions and divisions proposed by these authors, we perceive the following distinctions.

In the case of routine design,

• the design process "involves a well understood sequence of steps where all decision points and outcomes are known a priori" (Navinchandra, 1991, p. 2).

- "both the available decisions and the computational processes used to take those decisions are known prior to the commencement of a design" (Gero, in his Introduction to Navinchandra, 1991, p. vii).

In the case of nonroutine design,

- goals are ill-specified (D. Brown & Chandrasekaran, 1989).
- "the available decisions are not all known beforehand" (Gero, in his Introduction to Navinchandra, 1991, p. vii).
- one has neither fixed strategies nor preestablished decompositions and design plans for subproblems (D. Brown & Chandrasekaran, 1989).

In the case of innovative (conceptual) design,

- one "solves a known or a new problem in a way different from other known designs" (Navinchandra, 1991, p. 3).
- one produces designs outside the "well-defined state space of potential designs," by manipulating the applicable ranges of values for variables, attributing unfamiliar values to them. "What results is a design with a familiar structure but novel appearance" (Gero, 1990, p. 34).

In the case of creative design,

- one "uses new variables producing new types and, as a result, extending or moving the state space of potential designs. In the extreme case, a new and disjoint state space is produced" (Gero, 1990, p. 34).

The distinction that cognitive psychologists establish between routine and nonroutine problems refers to the knowledge that a person may bring to the corresponding task. According to Mayer's (1989) definitions of routine and nonroutine problems, which we adopted, the difference between these problems is the availability (and accessibility) in memory of a procedure allowing to perform the task.

Semantically Rich
Versus Semantically Impoverished Problems

This distinction is quite different from the previous ones. It was introduced in psychology and in AI. when more realistic and even real problems started to be tackled in these disciplines . It contrasts "semantically rich" (or "knowledge-rich") and "semantically impoverished" (or "knowledge-lean") problems (Bhaskar & Simon, 1977). In order to understand semantically rich problems, one needs specific domain knowledge for the construction of their representations. In order to solve them, one needs "strong" methods, that is, domain-specific problem-solving procedures. In order to solve semantically impoverished problems, one needs only, besides the information provided in the problem requirements, general knowledge and "weak," generally applicable methods, such as means–end analysis.

Design problems will generally be semantically rich—except if a design problem is to be solved by a problem solver without any knowledge in the domain. In this case, the problem situation is quite restricted, and thus impoverished—probably in an artificial way. Indeed, the distinction between semantically rich and semantically impoverished problems is again a relative one, which depends for a large part on the knowledge that a person may bring to a problem. Problems that are generally considered as semantically impoverished, such as artificial puzzles (e.g., the "missionaries and cannibals" problem), may become semantically rich for experts in that particular puzzle domain. Such experts may bring much specialized and diversified domain knowledge to a problem that for most people is exhaustively characterized by its specifications.

Interpretation Versus Generation Problems

This AI classification of problems is particularly relevant in a discussion of design problems.

B. Hayes-Roth et al. (1979) propose to distinguish generation problems from interpretation problems. B. Hayes-Roth and Hayes-Roth (1979) realized an exceptionally rich analysis and model of a particular instance of generation-problem solving, namely route planning, which is discussed later.

Interpretation problems are "problems which present the individual (or computer system) with the lowest level representation of the problem content (e.g., the speech signal) and require interpretation of the highest level representation (e.g., the meaning)" (B. Hayes-Roth et al., 1979, p. 382). Generation problems are "problems which present the highest level representation (e.g., the goal) and require generation of the lowest level representation (e.g., the sequence of intended actions)" (B. Hayes-Roth et al., 1979, p. 382). The authors comment:

> Interpretation and generation problems differ in important ways. For example, interpretation problems lend themselves well to initial bottom-up strategies, while generation problems lend themselves well to initial top-down strategies. Interpretation problems generally permit only one (or a small number) of solutions, while generation problems permit an arbitrary number of different solutions. Further, interpretation problems typically have correct solutions, while the correctness of solutions to generation problems varies under different evaluation criteria. (p. 382)

According to these distinctions, design problems are clearly generation problems. In addition to the definition, which applies perfectly to design problems, the further qualification proposed by the authors presents at least two other characteristics that we consider central in design problems: A problem has various, different solutions, and these solutions are not "correct" or "incorrect," but dependent on the criteria adopted for their evaluation.

Analysis Versus Synthesis Problems

A second classification developed in AI that seems interesting in the present context is the one presented by, for example, Clancey (1985) and the developers of KADS (see Breuker et al., 1987).

This classification is one of problem-solving tasks (and not of what we qualify as problem-solving *activities*). Clancey bases it on the concept of "systems" and their characterization in terms of inputs and outputs. Along these lines, he distinguishes synthesis and analysis as the two large sets of "generic operations" that may "do things to a system": Operations can construct a new system (synthesis) or interpret an existing system (analysis). Identifying design as a construction, that is, a synthesis task, Clancey distinguishes two types of design tasks: Configuration (characterizing a structure) and planning (characterizing a process).

The KADS knowledge elicitation methodology (Breuker et al., 1987) also adopts the top-level distinction between analysis and synthesis tasks, with modification tasks a transition area between the two. According to the type of input or output to a task, a finer classification can be established.

For synthesis tasks, the KADS authors adopt a two-stage model: (a) A stage in which an informal problem statement is being analyzed, which results in a formal specification of the structure to be synthesized, and (b) a stage in which this structure is effectively being created. Four types of design are distinguished depending on task input and output: Hierarchical, transformational, incremental, and multistream design.

The distinction between analysis and synthesis problems may be related to the ASE paradigm that underlies various engineering-design methods. In these methods, design is, however, not either an analysis or a synthesis problem, but involves analysis *and* synthesis, in two consecutive stages (cf. the two stages in synthesis that KADS distinguishes).

18.2.   ILL-DEFINEDNESS

As underlined already, we want to place a different emphasis on the ill-defined character of design, compared to Simon (1973/1984), but not only to him.

Simon (1973/1984) used the terms "ill-*structured*" and "well-*structured*." Instead of a problem's "structuredness," we refer to its "definedness," in order to encompass both the more or less well-structured, and the more or less well-specified character of all three problem components: Initial state, goal state, and operators. This position is inspired by Reitman (1964).

The notion "ill-defined problem" (or "ill-structured problem") is often presented as a synonym of "design problem." We consider both that ill-definedness is not entirely characterizing design problems and that design problems are not the only ill-defined problems. Other examples of ill-defined problems that have been analyzed in these terms can be found, for example, in management (Newell, 1969), judgmental and decision-making activities (Reitman, 1964), and in the social sciences, such as in political science (Voss et al., 1983).

The famous analogical-reasoning problems studied by Gick and Holyoak (1980, 1983) are considered "ill-defined insight problems" by their authors. Holyoak (1984) qualifies them as "ill-defined" "in that the permissible operations that might be used to achieve the goal are left open-ended. As a result, it is not immediately obvious how to apply a means–ends strategy, making it more likely that an available base analog will be perceived as useful" (p. 214). A consequence is that these problems allow "a variety of potential solution plans" (p. 215). Ill-definedness is thus not only applicable to problems that are solved in the real world: In spite of their real-world coverstories (a military fortress attack-by-dispersion, and a medical-radiation problem), Gick and Holyoak's problem-solving setting is the psychology laboratory.

With regard to design problems, their ill-definedness is a dimension: At one extreme, one may find exceptionally ill defined social-planning problems (such as identified by Simon, 1969/1999b) and, at the other extreme, design tasks that globally constitute routine problems, but that contain nonroutine subproblems requiring design activity: These are standard problems (e.g., in architecture, row houses). In between one may find innovative design problems (e.g., in engineering design, bridges with novel structures).

Furthermore, although our definition of ill-definedness is very broad, we consider that, in addition to their ill-defined character, design problems have other important features that are not determined by their ill-definedness (especially, their complexity, the role of constraints, the creativity involved in designing, the role of representation and knowledge, the need for collaboration between various specialisms).

The notion originally proposed by McCarthy (1956, as presented by Newell & Simon, 1972, p. 73) was a "well-defined problem." Newell and Simon refer to McCarthy when they declare, "a problem proposed to an information processing system is *well defined* if a test exists, performable by the system, that will determine whether an object proposed as a solution is in fact a solution" (p. 73). The authors add that they take "performable" to mean "performable with a relatively small amount of processing effort" (p. 73).

Minsky (1961) also, in his paper "Steps Toward Artificial Intelligence," only mentions "well-defined" problems, and proposes a comparable definition: We have to do with a well-defined problem, if "with each problem we are given some systematic way to decide when a proposed solution is acceptable" (p. 9). Minsky thus exclusively refers to a problem's solution and its evaluation procedure.

Eastman (1969) "extends . . . the information processing theory of problem solving . . . to include ill-defined problems" (p. 669; see also Eastman, 1970). Ill-defined problems are problems that lack (a) part of the problem specification and (b) a formal representation language. "Most such problems also lack a precise formulation of an acceptable goal state" (p. 669). In addition, many criteria that the specification must satisfy are left implicit (p. 670). "Ill-defined problems are subjectively specified" (p. 669). Eastman (1969) agrees, however, with Simon that "the search processes used by humans to solve both [well-defined and ill-defined] problems [are] similar" (p. 669).

In a case study of architectural design, Eastman (1969, 1970) examines a small-scale space-planning problem that consists of the selection and arrangement of elements in an existing bathroom that is to be redesigned. Even if this problem is not extremely ill defined, Eastman presents interesting observations concerning ill-defined problem solving. The author, for example, asserts, "the significant difference between well- and ill-defined problem solving is . . . a specification process similar to information retrieval processes" (p. 669).

Starting with Reitman (1964), design problems have been qualified as ill-defined because of the ill-defined character of all three problem components. Reitman was one of the first authors to provide an extensive discussion of the question (see also Voss & Post, 1988). Adopting an SIP approach, he distinguished different types of problems depending on which of the three components was specified more or less. Generally, in design problems, the goal state is the only component to receive some explicit specification, even if usually inadequately and at an abstract level, by way of the artifact's function and of other constraints on the artifact. Reitman applied his analysis to the composition of a fugue (Reitman, 1965, referred to in Reitman, 1964).

For Reitman, the concept of an "ill-defined problem" rests on that of an "open constraint" (Reitman, 1964, pp. 292–293), which is an attribute that is "open," that is, whose value is left unspecified (p. 314). Reitman indeed analyzes problem attributes as constraints on the problem.

In contrast to the classical cognitive-psychology typology, which distinguishes transformation from design or arrangement problems, Reitman (1964) takes together problems involving "transformation or creation of states, objects or collections of objects" (p. 284). He does so based on their equivalence with respect to the degree of specification of the different problem components.

Usually, compared to a problem's poorly, but somewhat specified goal state, the other two components, namely the initial state and operators, are extremely ill specified. In addition to the client's requirements and methodological norms, an important part of these problem components may be supposed to correspond to the state of the art in the domain and the relevant knowledge the designer possesses. These problem components may thus be supposed to be specified by means of these two sources of information. With respect to specific design projects, they are nevertheless clearly ill-specified.

It is by reference to Reitman (1964, 1965) that we consider design problems to be ill-defined with respect to all three problem-solving components — if one reasons in SIP problem-solving terms. This view has also been adopted by Thomas and Carroll (1979/1984), for whom "design is a type of problem-solving in which the *problem-solver* views his/her problem or acts as though there is some ill-definedness in the goals, initial conditions, or allowable transformations" (p. 222).

The definedness of a problem is thus a relative characteristic. Definedness, indeed, depends on the data available concerning the task, that is, on the state of the art in the domain under study, but especially on the knowledge possessed by the problem solver (covering also the problem solver's knowledge of this state of the art). As noticed by Schön (1988), "different designers construe the task they are asked to perform in very different ways, and their different readings of the task lead them to very different global patterns of designing" (p. 184).

Newell, Simon's SIP companion, also considers problem definedness a relative characteristic (1969). It depends on a problem solver's knowledge of problem-solving methods. "A problem solver finds a problem ill-structured if the power of his methods that are applicable to the problem lies below a certain threshold" (Newell, 1969, p. 375). A particular design brief may "look" ill-defined to a designer who "has only [his or her] general problem-solving abilities to fall back on" (Newell, 1969, p. 375). These same design specifications may look well-defined to other designers, whose experience or further knowledge leads them to evoke procedures for dealing with these specifications.

Reitman (1964) emphasizes that the distinction between well- and ill-defined problems is not simply the distinction between the formal and the empirical. There are empirical problems whose open constraints are "so limited in extent and importance as to make it reasonable to treat such problems (not merely formal models of the problems) as well defined for all practical purposes" (p. 308)—and artificially restricted problems that are nevertheless ill-defined (e.g., Gick & Holyoak's, 1980 and 1983, insight problems). An example is a jigsaw puzzle. We see again the continuum that ranges from well-defined formal problems to such ill-defined empirical problems as composing a fugue.

One might claim that, from a cognitive-activity viewpoint, most or all ill-defined problems might be analyzed as design problems—a perspective that we have suggested in the past (Visser, 1993b). Falzon (2004) even proposes to adopt design as a paradigm for analyzing all problem-solving activities. Eventually, Falzon posits, each design problem becomes a state-transformation problem, because of people's learning and acquisition of expertise, people's habits, and technological evolution. Falzon nevertheless also notes the possibility that there will always remain multiple possible perspectives and situations in which people refuse themselves to refer to procedures and routines. As an example, he refers to a study by Lebahar concerning painters who try to establish conditions that rule out the possibility to refer to routines.

Related Notions

Many authors have qualified design by reference to qualifications related to its ill-defined character: Wicked (Buchanan, 1990, 1992; Coyne, 2005; Rittel & Webber, 1973/1984), open-ended (D. Brown & Chandrasekaran, 1989; Dym & Little, 1999; Holyoak, 1984), real-world (Thomas, 1989), and vague (Newell, 1969, pp. 411–412).

Wicked Problems

Even if the notion "wicked" is often considered a synonym of "ill-defined," some authors make a distinction. Saunders (2001), for example, considers wicked problems a subclass of ill-defined problems. Sim and Duffy (2003) also distinguish the two, based on a Simonesque view of ill-structured problems. For Sim and Duffy, wicked problems "have no definitive or exhaustive formulation" whereas "ill-structured problems . . . become well-structured with decomposition" (p. 204).

The paternity of the notion is generally attributed to Rittel. In their discussion of "planning societal problems," Rittel and Webber (1973/1984, p. 136) qualify these problems as "inherently wicked," and oppose them to "tame" or "benign" problems. The authors oppose wicked social- or policy-planning problems to tame problems in the natural sciences. Because such planning problems depend on judgments—on political and other subjective views—, they are never completely or once and forever "solved." Another characteristic is that "every wicked problem is essentially unique" (Rittel & Webber, 1973/1984, p. 141). There are no classes of wicked problems in the sense that principles of solution can be developed to fit all members of a class. Despite seeming similarities, one can never be certain that the particulars of a problem do not override its commonalities with other problems already handled.

In a revisitation of Rittel and Weber's wicked problems, Coyne (2005, p. 12) proposes to "go further" than the authors did in 1973. For Coyne, "wickedness is the norm. It is tame formulations of professional analysis that stand out as a deviation" (p. 12). He relates the notion to recent theories about rationality and professionalism. "Perhaps the most provocative challenge comes from Deleuze and Guattari's difficult commentary on 'the rhizome,' which," according to Coyne, "has currency within much design studio culture" (p. 5).

Somewhat surprisingly, in our view, Coyne (2005) considers "controversial" his "conclusion that 'wickedness' is not aberrant" in design practice (p. 5). Coyne's conclusion is of course controversial by reference to classical psychological problem-solving models or prescriptive design models. In recent cognitive design studies, however, other views on design have been proposed. If the design one aims to cover is the activity implemented by professional designers, we consider "aberrant" to take as the reference such classical psychological or prescriptive approaches and views. The classical psychological models are based on data collected in artificially restricted contexts, such as the psychological laboratory. Prescriptive models do not cover design as it *is* (design practice), but as it *should be* implemented (corresponding to the *prescribed* design tasks that ergonomics studies continually show to not correspond to people's *actual* activity).

Vague Information

As noticed by several authors (e.g., Newell, 1969) "vague" is itself a vague term. Newell (p. 411) adopts, however, the notion of "vague information" in order to point to the important role that it plays in the distinction between ill- and well-structured problems. Newell uses the term in the standard sense of "ambiguous," namely "interpretable in two or more distinct ways" (Stacey & Eckert, 2003, p. 153). In their discussion of "ambiguity," Stacey and Eckert consider that the term "vagueness" is "best reserved for the failure (to some degree) of a representation to enable a sufficiently clear and certain interpretation" (p. 167). Representations such as sketches are ambiguous rather than vague, the authors claim, "when alternative ascriptions of symbols to [their] elements are possible" (Stacey & Eckert, 2003, p. 171).

Design Characteristics
Associated With Its Ill-Definedness

We conclude this section on design problems' ill-definedness with a brief discussion of three related aspects.

A factor contributing to design problems' ill-defined character, but that is seldom referred to in the research literature in this context, is the important role of clients' and prospective users' involvement in requirements development (Carroll, 2000). This factor affects the ill-definedness of both design problems' initial state (the artifact's requirements evolve in a requirements development process; see Carroll et al., 1997, 1998) and their goal state (by consequence, its purpose and functions also evolve). Moreover, one may suppose that ordinary, lay persons such as clients or users adopt other procedures in their design contribution than the "official" designers and other "technical" participants.

The ill-definedness of design is, at least in part, at the source of its satisficing character. The impossibility of optimization depends undoubtedly on the ill-definedness of design problems' goal state, but is also related to that of the other two problem components. Given the ill-definedness of its initial state, a problem can be interpreted and thus be solved in various directions, and admit so many alternative solutions that only satisficing, no optimizing, is possible (the viewpoint adopted by Simon).

The ill-defined character of design problem operators has consequences at different levels for the solving of design problems. First, at a local level, even if designers have many preestablished design methods available, this repertoire is not necessarily sufficient. Designers often have to import procedures from other domains, both technical and commonsense knowledge. In our carrying/fastening device study (see Table 6.1), for example, we showed the importance of designers using commonsense knowledge in industrial design (Visser, 1995b). In addition, even if at a local level designers use preestablished strategies, the articulation of strategies will not be preestablished at the organizational, that is, global, level of their activity (such as top-down, breadth-first proceeding). Designing has an opportunistic organization (discussed later).

18.3.   COMPLEXITY

Design problems are often qualified as "complex" because they are large and require many competencies. In our view, more essential than their size is their complexity as defined in cybernetics, where a system's complexity depends not as much on the number or the density of its elements, as on the relations between them, their pattern, and the unforeseeability of the types of relations.

In design problems, the interdependencies between the generally large numbers of components are indeed often difficult to anticipate. To make resolution more feasible, a definite decomposition of the problem would be necessary. Usually, such decomposition is, however, hard to accomplish.

18.4.   CONSTRAINTS ON THE ARTIFACT

Several authors, especially in AI, analyze and model design as management and satisfaction of constraints (Feitelson & Stefik, 1977; Logan & Smithers, 1993; Stefik, 1981a, 1981b; Steinberg, 1987).

Cognitive ergonomists have examined if designers indeed use constraints—and if they do, how they proceed. Bonnardel (1989, 1992) has analyzed the exploitation of constraints in design evaluation, in the domain of aerospace composite-structure design. Darses (1994) has examined constraint management in studies on design development in the domain of local area networks. Applying a cognitive-ergonomics viewpoint on AI constraint-based approaches to design, Darses (1990a, 1990c) concludes that, even if these interdependent variables play a central role in design, designers' activity also refers to other types of knowledge, such as action plans and schemata.

Formally it is possible in most cases to transform an ill-defined problem into a well-defined one by closing all open constraints (which are for Reitman, 1964, the most important factor of design problems' ill-definedness). Such transformation corresponds, however, to a form of premature commitment. A designer can restrict the solution space by selecting, right from the start, a particular concept. Yet, designers may well regret afterward such premature commitment to a selected "kernel idea." Early constraining of variables will frequently lead to unsatisfactory design solutions.

Introducing additional constraints can narrow down the space of solutions. Adding constraints into a task is a process that is not specific to design, not even to problem solving. For a long time, now, ergonomists have noted the tendency of people to constrain their task more than prescribed.

Reitman (1964) notes that the composer whom he observed "quite explicitly leaves open the use to be made of the countermaterial" corresponding to an open constraint (p. 293). "Frequently it is just exactly the openness of certain constraints that makes a solution of the problem possible. These constraints provide definitional stack" (p. 293).

Most SIP authors who have discussed constraints judge that, among the three components of a problem, its goal state is especially constrained. Therefore, in order to guide their activity, designers would start with the constraints that govern the problem's goal state. These constraints would guide their narrowing down the set of possible solutions (Greeno & Simon, 1988). Simon (1992a) notes that "one of the skills that the professional designer acquires in any domain is discernment of which constraints to satisfy first so that the plan can be carried to completion with a minimum of revision" (p. 134).

In his analysis of design in terms of satisficing, Simon focuses on evaluation. We consider that, in addition to their "critical" function (as evaluation criteria), constraints may play a generative role as well (Visser, 1996). It is only in the context of particular types of design, such as social planning, that Simon also considers generative constraints (e.g., "interestingness" or "novelty").

The constraints that design problems are to satisfy are often conflicting. Dealing with trade-offs among constraints, and with different possibilities of how to do so, is typically a matter of satisficing. In combination with the complexity of design, that is, the multidimensional and interdependent organization of its variables, this leads to making the analysis, structuring, and pruning of constraints into an essential aspect of design activity.

Many constraints are open to discussion—certain types of constraints more than others: Particularly open to debate are social and political aspects of an artifact, whereas financial aspects are probably less negotiable—and legal aspects even less (Rittel, 1972/1984, p. 325). There are, however, also constraints that cannot be discussed at all, such as nomological constraints, which are dictated by natural law—they cannot be violated.

Flexible or not, constraints are adapted and modified by designers. Goel (1995) qualifies as "reversing the direction of transformation function" the fact that designers may "occasionally stop and explicitly try to change the problem parameters by manipulating both the problem constraints and the client's expectations" (p. 101). "Since the task structure is not well specified in advance and the constraints are nonlogical, the designer can negotiate, narrow, or simply change problem parameters" (p. 92).

Akin (1978, quoted in Cross, 2001a, p. 82) also observes that "one of the unique aspects of design behavior is the constant generation of new task goals and redefinition of task constraints."

According to a commonsense belief, constraints are often considered as obstructing people's "freedom," and thus—one reasons—their creativity. However, constraints are useful, if not necessary, in narrowing down a space of possibilities that can otherwise be too large for search or exploration. Something that is also counterintuitive relative to common sense is that expression of artistic creativity requires constraints. Creating without any constraint is extremely difficult! As claimed by Stacey and Eckert (2003) on the basis of empirical research, "hard constraints foster both creative designs and the development of flexible procedures for developing innovative designs" (p. 166).

Analyzing the ill-defined problems involved in composing a fugue, Reitman (1964) notices that "though they would generally be considered complex they include few constraints as given" (p. 296). In the light of the current discussion, their unconstrained character could thus exactly be one of the reasons for their difficulty!

Reitman (1964) proposes different sources of constraint propagation:

• Use "transformational formulas": Applying structural or syntactic con- straints, a designer transforms one constraint into several others by way of decomposition (e.g., "fugue" -> "exposition" + "development" + "conclu- sion").

• Increasingly specify subcomponents and, doing so, immediately further particularize constraints on any other subcomponents that may have been defined in relation to them.

• Adopt conventions for the choices and decisions one has to make.

• Proceed to "curve fitting," that is, after-the-fact adoption of a convention consistent with what already has been done.

"Conventions" can be defined before the generation of the material to be gov- erned by them, or adopted after-the-fact (two figures also observed in relation to the search for homogeneous representations by the software engineer we ob- served; Visser, 1987b). Reitman (1964) notes that often a conflict arises "be- tween the requirements of unity and economy in the use of conventions and the need to maintain interest and variety within particular subcomponents" (p. 299). In our opinion, this indeed holds in domains in which "interest" and "variety" are relevant evaluation criteria for the artifact (e.g., music composition, architecture, and other types of design not exclusively governed by technical constraints), but much less, or even not at all, in technical domains where such evaluation criteria are not positively valued in particular.

In the protocol analyzed by Reitman, the composer is observed to spend more and more time characterizing what he already has. He does so in order to redeem loosely stated constraints requiring further particularization. His context is in- creasingly well-specified and he has to proceed by reference to the details of this context.

Temporal and Spatial Constraints

Designing plans (one form of planning, besides plan execution) is a design activ- ity in which temporal constraints clearly play a central role. Several early em- pirical design studies have been examining these constraints in the context of planning but also in other design activities. Some planning tasks combine them with other types of constraints. Route planning is a design activity that has the particularity to combine temporal and spatial constraints. The famous B. Hayes-Roth and Hayes-Roth (1979) study has examined this type of design (for a comparable, more recent study, see Chalmé et al., 2004).

Other studies have compared spatial design with temporal design. Carroll, Thomas, Miller, and Friedman (1980) used an experimental approach to make this comparison, using spatial and temporal design problem isomorphs. The spatial form of the problem led to more adequate solutions than its temporal isomorph, and the spatial-problem solutions were designed in less time. This difference vanished when the authors provided participants with representation aids. Thomas and Carroll (1979/1984) conclude that there is "a notably strong tendency for spatial problem statements to encourage the use of graphic representational aids that were not encouraged by the temporal problem statement" (p. 232).

One might be inclined to conclude that temporal constraints are more difficult to handle in design than spatial constraints. A more recent series of studies shows that things are probably more complex.

In the context of a research project on spatio-temporal cognition (Visser, 2002b; Visser & Wolff, 2003a, 2003b), Chalmé has examined the design of route plans (Chalmé et al., 2004). Route planning, by definition, imposes spatial constraints right from the start: Items to be organized—that is, ordered on a time axis—into a route plan are specified to be at particular locations in space. In addition to the implicit temporal constraint present in all planning tasks, route planning may contain additional, explicitly formulated temporal constraints, for example, that a particular action has to be performed, or a location reached, before a certain hour.

A first experimental route-planning task specified a series of chores in terms of both types of constraints, which therefore were to be integrated in a spatio-temporal route plan. Depending on their spatial knowledge—here, knowledge of the environment to be traversed—people performed this constraints articulation differently (Chalmé, Visser, & Denis, 2000; Chalmé et al., 2004):

- People with spatial knowledge of the environment took into consideration the spatial aspects of the task before its temporal aspects, whereas people without this spatial knowledge took into consideration the temporal aspects before the spatial ones—as if confronted with a "simple" planning task.

- People with spatial knowledge tried to optimize the duration of the itinerary, whereas people without this knowledge tried to optimize its length.

A second route-planning task—involving different participants than the first study—focused only on people who beforehand had no spatial knowledge of the environment. Conditions were less constrained: (a) Participants were provided with additional spatial information about the road network that seemed to have been useful to their knowledgeable colleagues during the first task, and (b) there were no explicitly temporally constrained chores. In this situation, these people without spatial knowledge of the environment tended to plan like their knowledgeable colleagues, with respect to the differential processing of spatial and temporal aspects. Integrating the spatial and temporal aspects of the tasks, however, continued to be difficult for these amended "novices."

These studies concur thus with previous research, in showing that spatial and temporal constraints are processed differently. In addition, they seem to indicate that

- the status of these two types of constraints varies with people's spatial knowledge of the environment to be traversed.
- this disparity is not uniform for spatial and temporal constraints:
  - o People with spatial knowledge favor spatial aspects with respect to the order in which they handle the constraints pending on a route, but they favor the temporal aspects as a route-optimization criterion.
  - o People without the corresponding spatial knowledge proceed in the opposite way: They first handle the explicit temporal constraints, but optimize the route on spatial criteria.
- the integration of both types of aspects is apparently difficult: It seems to call for good spatial knowledge, in any case knowledge of the environment to be traversed. Providing people who do not posses such knowledge with the corresponding spatial information does not necessarily enable them to proceed to such integration.

We presented earlier the classical result with respect to constraint processing often identified in ergonomics studies: Given a task, people, in their actual task, generally *add* constraints to those imposed by the prescribed task. Several examples were noticed in the route-planning studies.

To start with, there was the case of implicit temporal constraints (ice cream and flowers to be bought whenever one wishes to), whose inclusion in the route plan might be considered as adding constraints. A more interesting example was the case of the constraint that the route be of the shortest length possible. This spatial constraint, which had not at all been prescribed, seems to have been added and applied by many participants. Indeed, many participants tried to minimize the distance between chore locations—and thus the total route length—by applying a "nearest-neighbor" strategy. This strategy, however, seems to have been applied less widely than in "traveling salesperson" problems that presented this spatial constraint explicitly as such (Best & Simon, 2000).

At first view, planners might consider the planning of a route between chores without explicit temporal constraints, as an *implicit* traveling-salesperson problem. In our route-planning studies, people did not always approach their task in this way. This may be due to several reasons:

- Contrary to what characterizes traveling-salesperson problems, the minimal-length constraint was presented neither explicitly nor implicitly (insofar as implicitly presented information can be controlled).

- Being faced with a real city (even if only on a map), participants may add constraints to their prescribed planning task in order to elaborate a realistic route, by incorporating plausible temporal task attributes (e.g., consider that a theater booking office does not open until late in the morning).

- Best and Simon (2000) note that traveling-salesperson problem solving may be best explained by planners adopting a meta-strategy that combines global and local strategies (especially, in the form of the nearest-neighbor strategy). In one of our route-planning studies, such meta-strategies combining global and local strategies (nearest-neighbor and other strategies) were observed among participants who were familiar with the environment to be traversed. The participants who did not know the environment may have come to adopt a comparable approach when their mental load was alleviated (i.e., when there were no explicit temporal constraints pending on the chores in their route-planning task).

All questions concerning the relative difficulty of processing spatial and temporal constraints, and their articulation and possible integration are thus clearly not yet elucidated.

19 Intermediate Representations

Intermediate, transient representations are the representations that clearly occupy the greatest part of the design activity during a project. The representations produced and used in early and later intermediate design phases are generally not of the same type as the final representation, which specifies the implementation of the artifact. They allow designers to focus on different aspects of their design (Newman & Landay, 2000), which may or may not be maintained until the final design stages. Indeed, intermediate representations are observed to contain information not found in the final representation. Moreover, as shown by J. C. Tang (1991) in his research on collaborative work in conceptual design team sessions, "the *process* of creating and using drawings" also conveys important information not contained in the resulting drawings (p. 156; the emphasis is ours).

In addition to being *intermediate* between the requirements at the start of a design project and the specifications at its other extremity, representations can also have an *intermediary* function. Two types of intermediary representations can be distinguished: Those between designers and the object of their activity, and those between several designers.

Both functions are discussed in chapter 21, on "Designing as an Activity: Construction of Representations."

19.1.   THE EVOLVING NATURE
OF REPRESENTATIONS:
LEVELS OF ABSTRACTION

One of the first results with respect to representations in cognitive design research conducted in architecture, was that representations constructed and used in early phases differ from those constructed and used later on (Lebahar, 1983).

An important difference concerns two orthogonal types of abstraction levels: Implementation hierarchy (Rasmussen's, 1986, "abstraction hierarchy") and part–whole hierarchy (aggregation hierarchy). With respect to the first dimension, representations differ in concretization, from those related to the artifact product's purpose or function, to those related to its physical structure and other properties that specify its implementation. As regards the second dimension, representations differ from global to detailed (granularity).

Notice that we do not use "refinement" for the part–whole hierarchy, as authors often do, using the term to refer to both "concretization" and "detailing."

In their activity, designers do not progress through these two hierarchies in a systematic, fixed order, traversing them from abstract to concrete and from global to detailed. Instead, they come and go between representations at level n and level n±m in each one of these two hierarchies. Many authors (e.g., Goel,

1995) present design as a quite systematic process proceeding in different, con-secutive stages—a view that we attribute to their data collection in a laboratory context. In his studies, Goel (1995) observes proportionally more activity at a functional level in early stages and more activity at a structural level in later stages (cf. his design-development categories and aspects). He also observes that early on in the project, the artifact product will be specified in less detail than in later stages. *Globally*, these tendencies also hold for professional design activi-ties. We see the differences and other details later on.

19.2. THE EVOLVING NATURE OF REPRESENTATIONS: DEGREES OF PRECISION

A further difference between representations in early and in later phases is that initial representations are necessarily rather imprecise (Goel, 1995; Lebahar, 1983; cf. Stacey & Eckert's, 2003, analysis in terms of "ambiguity"). This may be due to the information possessed by designers about their design project not being enough in either quantity or precision. However, another, more essential factor is that, in early design, designers need flexible, fluid forms of representa-tion that, for one thing, express and convey the provisional character of the un-derlying idea, and for another, prevent the designers from premature commit-ment to specific options. They need representations that allow them to maintain as many degrees of freedom as possible in their evolving design. The more de-tailed and the less ambiguous a drawing is, the more it constrains interpretations (Tweed, 1999). Only gradually, as design progresses, are initial representations translated into representations with increasing degrees of precision and that, for example, in domains where external representations such as drawings are greatly used, are rendered in other types of drawings (McGown et al., 1998).

In such external representations as drawings, these qualities of provisionality and not-fixedness may be translated through roughness, which is why sketches are so useful and frequent in the early stages of design in domains such as archi-tecture (Goel, 1995). Using observations of sketching by mechanical engineer-ing students, during a 15-week period of conceptual design, McGown et al. (1998) note that "the sketching activity has peaks and troughs over time, with its highest peak near the beginning" (p. 452).

Notice that it is not always obvious to separate in external representations the implementation of decreasing levels of abstraction and of increasing levels of precision. Detailing a representation generally requires more precision. The distinction between the two types of levels has above all a heuristic function. It allows, for example, distinguishing designers' activities that are more oriented toward communication and interaction with their colleagues, from activities that do not especially possess this orientation (but aim, e.g., generation of a new solution, without consideration of its eventual, later submission to one's colleagues).

20 Representations at the End of a Design Project: Specifications and "Design Solutions"

The ultimate representation constructed by designers during their work on a design project is supposed to express the artifact's specifications—or, in problem-solving terms, the solution to the problem. Three aspects of an artifact must be expressed by these specifications: *What*—the artifact product itself—*how*—the procedure by which it should be implemented—and *why*, that is, "the reason why the design should be as it is" (Stacey & Eckert, 2003, p. 170).

The *why* is often neglected—or even omitted—in design documents. For some 15 years now, work on design rationale has emphasized its importance (Buckingham Shum & Hammond, 1994; Moran & Carroll, 1996a). The term "design rationale" is used in many different senses, but what is central in most approaches is the capture of the reasons behind design decisions and the idea of justification (Moran & Carroll, 1996b).

In this chapter, we discuss two characteristics of these ultimate design representations. The first characteristic, namely the fact that design-problem tasks may lead to several satisfactory solutions, has been mentioned in most cognitive design studies. The impossibility to test the artifact product specified by the solution has not been the object of much discussion. We close this chapter by a presentation of the rare cognitive design studies that have been examining the quality of design solutions.

20.1. SEVERAL SATISFACTORY SOLUTIONS INSTEAD OF ONE CORRECT SOLUTION

On the one hand, different designs may satisfy the same requirements. On the other hand, "no solution to an ill-defined problem can count on universal acceptance" (Reitman, 1964, p. 302). This characteristic is related to design problems' ill-definedness and to the satisficing nature of the design activity. The different solutions that may result of a design process are indeed clearly related to the various approaches that are possible when different designers represent their task and proceed in this process.

The solution formulated by a designer is, generally, neither "correct" nor "incorrect." It may be more or less satisfactory compared to one or more other acceptable solutions, which may differ more or less on several characteristics and with respect to several criteria. A first solution may be judged "better" than a second one for reasons of implementation; another may be preferred for reasons of maintenance; a third seems better because of user-oriented criteria, and so on (Visser & Hoc, 1990).

The final specifications for an artifact product proposed by a designer may thus be contested by a colleague or—what is worse—by the client, not as much for being incorrect, but because they adopt other criteria than the author. This eventuality indicates that, as for solution proposals, design criteria cannot be ordered according to an objective importance or validity rank, unanimously accepted by all participants involved in the design project. In our study of software-review meetings, we indeed observed how designers discuss criteria (D'Astous et al., 2004). Contrary to what holds for well-defined problems, there exists for ill-defined problems neither "a definite criterion for testing any proposed solution, [nor] a mechanizable process for applying the criterion" (cf. Simon's, 1973/1984, list of criteria for well-structured problems).

Different designers come up with different design proposals—that is, more than one idea is formulated per designer, and designers formulate ideas that are different from those advanced by their colleagues. This has been observed in various domains, for example, architecture (Eastman, 1970), mechanical design (Frankenberger & Badke-Schaub, 1999), software design (Malhotra et al., 1980), and traffic-signal setting (Bisseret et al., 1988). Goel (1995) observes the use of both institutionalized and personalized stopping rules and evaluation functions (pp. 99–100).

In software design, "there is no way to determine which design or program code is better than another" (Kitchenham & Carn, 1990, p. 275). One of the underlying factors is that different ideas about the users' future work with the system under design can lead to different design solutions, one of which cannot be considered better than another (Löwgren, 1995).

In their discussion of wicked societal-problems planning, Rittel and Webber (1973/1984) have presented 10 "distinguishing properties" of wicked problems. At least half of them are related to this characteristic that links design problems and their solutions in a rather loose way.

The first distinguishing property is that "there is no definitive formulation of a wicked problem" (Rittel & Webber, 1973/1984, p. 136). Each formulation of a wicked problem corresponds to the formulation of at least one solution—we would say: Each representation of a wicked problem corresponds to the representation of at least one solution. In Rittel and Webber's terms, problem understanding and problem resolution are concomitant to each other.

The second property is that "wicked problems have no stopping rule" (Rittel & Webber, 1973/1984, p. 138). This is mainly because there are no criteria for "sufficient" understanding and because there are no ends to the causal chains that link interacting open systems. Designers stop because of considerations external to the problem: They run out of time, money, or patience.

"Solutions to wicked problems are not true-or-false, but good-or-bad" (p. 138) is the third property enunciated by Rittel and Webber (1973/1984).

Another property that is clearly applicable here is that "wicked problems do not have an enumerable (or an exhaustively describable) set of potential solutions, nor is there a well-described set of permissible operations that may be incorporated into the [artifact]" (p. 140).

Rittel and Webber (1973/1984) also state that "there is no immediate and no ultimate, definite test of a solution to a wicked problem" (p. 139). The consequences of a design solution extend over a long period and are of concern to many people. The recognition of these human-impact type of consequences by Rittel and Webber establishes, in our opinion, an aspect that was absent from the characteristic "absence of a definite test criterion" that Simon (1973/1984) listed as one of the features contributing to the ill-structuredness of a problem.

20.2. IMPOSSIBILITY OF TESTING THE ARTIFACT SPECIFIED BY THE SOLUTION

In addition to lacking a "definite" test criterion, an artifact product that is "under design," by definition, cannot be tested. In order to be tested, it has to have been implemented already. A consequence is that omissions, failures, and other trouble often go unnoticed until it is "too late." Another effect is that design never ends (see also Rittel and Webber's remarks on social-problem solving).

What designers can "test," or evaluate, are representations of the artifact product. This testing is more or less feasible, difficult, and conclusive, according to the domain of design.

Under the heading "Feedback Loop," Goel (1995) notes that "there is no genuine feedback from the world" during design. "Real-world feedback comes only after the design is completed and the artifact is constructed and allowed to function in its intended environment. At this point, the feedback cannot influence the current project, but only the next 'similar' project" (p. 86).

20.3. QUALITY OF DESIGN SOLUTIONS

It seems significant to us that cognitive design research has not been greatly occupied with the quality of design, neither with that of the design process, nor with that of its result (but see Fricke, 1992, 1999; see also Détienne, Burkhardt, & Visser, 2004; Rodgers et al., 2000; Van der Lugt, 2003). Researchers in this domain have noticed that an artifact product may take different forms without one being better than the other, but they have rarely examined cognitive factors underlying design quality. Noncognitive factors have been discussed by design engineers and methodologists, who have been concerned with measuring, or estimating, both the effort put into the process and the quality of the result (see Jedlitschka & Ciolkowski, 2004, for software engineering). We have defended elsewhere the idea that measuring process effort and product quality, and establishing a relation between the two cannot be performed without a model of the cognitive activities involved in the design task, and without a measurement of these activities (Détienne et al., 2004). Today, the data that cognitive models may provide regarding this issue is, however, still sporadic.

It is probably no coincidence that one of the rare cognitive design studies on this theme has been conducted in the framework of a collaboration project between psychologists and engineering-design researchers. In one of the "German empirical studies" on engineering design (Frankenberger & Badke-Schaub, 1999; Pahl, Badke-Schaub et al., 1999; Pahl, Frankenberger, & Badke-Schaub, 1999), Fricke (1999) has identified several characteristics of the activities of successful designers. The author has examined the possible consequences of different problem specifications ("varying levels of completeness and detail in the assignment") on goal analysis and search for solutions, and on the quality of the results, analyzing the design process in terms of Pahl and Beitz's stage model (1977, 1984). Nine designers who all had been taught design methodology were presented with "two differently formulated, but in principle identical problems" (p. 418). "Five designers were given an extensive and precise assignment. . . . Four designers were given the problem in an incomplete, imprecise formulation" (p. 418). "Against [the author's] expectations, . . . there was no relationship between the type of problem formulation [precise or imprecise] and the total design time, despite the differences in time used for the goal analysis. Neither could any relationships be found between the solution quality and . . . duration of the goal analysis" (Fricke, 1999, p. 422). There were two successful designers in each condition. Examination of the approaches used by successful designers led the author to conclude that successful designers

- clarified problem specifications, focusing on problem structure.
- proceeded to "critical analysis," "actively [searching] for information, critically [checking] given requirements and [questioning personal] requirements regarding their priority".
- "summarised information of the problem formulation into requirements and partially prioritised them".
- "did not suppress first solution ideas even in the goal analysis," but also "repeatedly tried to return to 'clarification of the problem' until this phase had been completed".
- "detached themselves during the conceptual design stage from the pre-fixation generated during the goal analysis".
- "produced variants, but kept an overview in that they alternately generated solutions (block-wise), and then assessed them consciously to reduce the number of suitable variants".
- performed "better technical assessment" than their colleagues, using the advanced engineering knowledge they possessed (pp. 428–429).

Based on this last point, Fricke (1999) the author remarks that design methodology "does not make up for a lack of engineering knowledge, but can assist in obtaining good solutions if it is applied flexibly in accordance with the problem to be solved" (p. 429). Considering that "the designers who produced good solutions used approaches that encourage success" (p. 428), Fricke uses these results to formulate recommendations for design methodology. To give an example: The observed detachment from specific solution ideas during the conceptual design stage leads Fricke to suggest that "solution-neutral formulation of requirements seems to be an important pre-requisite" for successful design.

One may notice that this "solution-neutral formulation" will generally have to be a reformulation by the designer—otherwise the recommendation would be at the address of the client.

We consider it relevant to identify the characteristics of successful designers' activities. Many empirical design studies, however, have shown at length that designers often do not implement the corresponding "success-giving" procedures or strategies. Therefore, a next step is to develop ways to make designers effectively apply these strategies. Given the history of the effective application of design methodology established in cognitive design research, this seems rather difficult! Another obstacle to effective implementation of these good approaches is that the application of design methods does not automatically produce better designs than nonsystematic design, as has also been shown in the German empirical studies mentioned earlier (Pahl, Badke-Schaub et al., 1999).

With respect to creativity, that is, a factor considered particularly essential in early conceptual design as "leading to original and useful products" (as creativity has been characterized), Van der Lugt (2000, 2002, 2003) has conducted several studies. In his research on creative problem-solving methods, he has examined various techniques that are supposed to structure and stimulate creativity. In one of his analyses, Van der Lugt (2003) studies the relationship between the creative qualities of ideas generated and further developed during design, and the integratedness of the design process with respect to those ideas (see Roozenburg & Dorst, 1999, for the qualification of "designing in an integrated manner" as beneficial). Van der Lugt's data supports the hypothesis that ideas with a strong network of connections are judged as more creative than ideas that lack such a network.[29]

Based on Goel's (1995) distinction between lateral and vertical transformations, Rodgers et al. (2000) propose the assessment of the lateral–vertical transformation balance as a method to improve the efficiency of sketching in design. According to Goel (1999), "a vertical transformation is one where movement is from one idea to a more detailed version of the same idea. It results in a *deepening of the problem space.*" "A lateral transformation is one where movement is from one idea to a slightly different idea rather than a more detailed version of the same idea. Lateral transformations are necessary for the *widening of the*

---

[29] The judgment was performed by the participants in the meeting in which the ideas were produced, that is, by the authors of the ideas and their colleagues, not by independent judges.

problem space and the exploration and development of kernel ideas." Rodgers et al. (2000) claim that in early, conceptual design, "good" design is characterized by a balance between lateral and vertical transformations "rather than an extreme *lateral* bias" (p. 461). "It is likely, however, that the balance will shift to an extreme (and finally total) vertical bias as the design representation progresses towards the embodiment and detailing stages" (p. 461).

In the engineering-design domain, Hubka and Eder (1987) consider that mainly four design factors affect the quality of design: The design engineer, the working means, the knowledge or know-how (methods, technical information, representation methods, management), and the working conditions and environment. For each factor, the authors indicate its influence on each of a number of design-process characteristics, namely quality of technical system, duration of design process, efficiency of design process, risk of failure, transparency of design process, nonroutine designing, cost of the design process, and quality of description (p. 132). They assert—without, however, any presentation of further underpinning—that design engineer and working means are most influential, whereas working conditions and environment have only a small or even no influence on the design process. This prediction seems rather surprising to a researcher in ergonomics, whose field is based on the idea that working conditions and environment are essential factors influencing people's activity—an idea that has been shown by and large to correspond to reality!

With respect to organizational factors influencing the quality of design, Winograd, in his series of essays *Bringing Design to Software* (1996), proposes Norman's famous "war story" on the complications of "something as simple as the placement of the power switch" on the consecutive exemplars of a series of Apple computers (as qualified by Swaine, 1996). Winograd introduces Norman's paper as an analysis of the kind of culture or organizational structure that promotes effective design. Norman (1996) reports in detail "how a dedicated committee tried to simplify the placement and function of the switch, but succeeded only through multiple compromises in the face of many reasonable technical problems" (p. 234). What Norman especially describes in great detail is how the Apple organizational culture makes design a complicated matter, even the design of something as simple as a power switch on an Apple. He presents the enormous number of compromises that must be reached between a big number of people who come from different backgrounds, and thus have different representations of the design and hold different stakes in it.

21 Designing as an Activity: Construction of Representations

This chapter presents the different activities involved in designing, that is, in the construction of representations of the artifact that is the object of design.

As mentioned in the discussion of Simon's SIP approach to design, we consider essential two of the definitional cognitive aspects of designing identified by Simon. These are the view of design as a satisficing activity and as a type of cognitive activity rather than a professional status. A short review of the first of these two design characteristics follows. The second characteristic is discussed in a later section.

A separate section discusses the relations between designing and the tasks surrounding it, namely requirements development and implementation.

Before introducing the different activities that make up design, we review the two higher levels of the global design activity (Visser, 1992b): Its organization and the more or less local strategies that professional designers implement, namely reuse, simulation, selection of a kernel idea, and guidance by user considerations.

An essential characteristic of designing is, in our opinion, that its organization is opportunistic. We discuss this position in some detail, because not all colleagues in cognitive design research share our view.

21.1.   DESIGNING AS SATISFICING, BOTH IN GENERATION AND IN EVALUATION OF SOLUTIONS

We introduce one nuance relative to Simon's analysis of design as a satisficing rather than an optimization activity. We consider that satisficing occurs not only in evaluation, but also in generation of solutions. Eastman (1970) observed these two forms of satisficing. Optimization was rare in his study: "Only one instance of an attempt to optimize a solution has been found in thirteen protocols concerning three different design tasks" (p. 30).

21.2. THE INTERMINGLED CHARACTER OF REQUIREMENTS DEVELOPMENT AND DESIGN, AND OF DESIGN AND IMPLEMENTATION

Requirements development (a term proposed by Carroll et al., 1997, 1998; traditionally called "requirements specification" or "requirements definition") refers to a task that is supposed to precede design.

According to a common viewpoint, requirements are "out there" and are simply to be "captured" or "gathered" at the beginning of a development project. In the course of a series of participatory design sessions, Carroll (1997, 1998) noticed, however, how project requirements evolved. "The client's original functional requirements . . . were radically and continuously transformed. . . . Qualitatively different requirements become accessible or salient" (Carroll et al., 1998, p. 1167) all through this process of collaborative requirements gathering that the authors qualified as "developmental" in the sense of Piaget, Inhelder, and Vygotksy (Caroll et al., 1998, p. 1156). Many of the new requirements were "nonfunctional requirements pertaining to workplace practice—a category absent from standard taxonomies of nonfunctional requirements" (Caroll et al., 1998, p. 1156). The authors did not analyze this as "initially mistaken notions being subsequently corrected, or [as] more requirements work leading to successively finer decompositions" (Caroll et al., 1997, p. 62). They qualified the activity as requirements "development" (see also Carroll, 2000; Papantonopoulos, 2004)—we would qualify it as *design* of requirements.

A comparable relation may be identified between design and implementation. The implementation (realization, manufacturing, fabrication, construction) of an artifact based on the specifications that result from the design, is also a different task from design, assigned to different professionals. Yet, it is generally interspersed with more or less important design activities, because neither the explicitness nor the completeness of the specifications is absolute. Even if the design project has been declared finished (the specifications—plans or other representations—are transmitted to the workshop or other construction departments), the artifact's implementation may still require design. People realizing the implementation activities will often have to take decisions that, from a cognitive viewpoint, constitute designing. "Design never ends." Later sections discuss other design prolongations, especially through maintenance and users' participation in design.

The reverse holds as well: During design, there are already implementation activities. In our software-design study, for example, we observed that specifying software and not coding it right away may be difficult for software designers (Visser, 1987b, 1992b) (see the following example).

Example. The interweaving of design and implementation (i.e., in this case, coding) was observed in our software-design study (see Table 6.1). We observed that the first day of work on this design task was dedicated to an initial, global understanding of the client's requirements. The software engineer analyzed (in fact, skimmed through) the documents he had received (a 50-cm pile of A4 papers). At the end of this first day, during 1 hour, he made a global plan for his coming designing of the IPC software. After this unique day of analysis and planning, that is, from the second day on, the software engineer directly started to code. This coding was often interrupted—of course, we would say—for one or the other of various other types of activities, from planning to design evaluation. These activities did not necessarily occur in iterative cycles that were executed in a fixed order. Thus, the software engineer's interruptions of the coding activity were not systematic. That is why his activity during the rest of the 4 weeks that he took for developing the software could not be further decomposed into functionally homogeneous phases. We concluded that design took place during the entire development process, completely intermingled with implementation of the resulting design choices (Visser, 1987b). Carroll (2000) speaks of the "actively synthetic design method of planning by doing that is complementary to the relatively analytic techniques of problem structuring and decomposition" (p. 29).

Notice that another interpretation of these observations would be that the software engineer completed his entire design in only one hour, implying that there was nearly no design taking place—the software engineer's activity consisting of directly and almost exclusively coding the concept that he designed in 1 hour. We do not adhere to this alternative interpretation.

21.3. DESIGNING
AS A COGNITIVE ACTIVITY
RATHER THAN A PROFESSIONAL STATUS

We analyze design as a type of cognitive activity, not as a professional status (idea advanced by Simon, 1969/1999b). This means that the activities of many professionals who are not categorized as "designers," would be qualified "design." Simon (1969/1999b) proposed a very broad definition of design when he wrote that "everyone designs who devises courses of action aimed at changing existing situations into preferred ones" (p. 111). We do not consider as designers physicians prescribing remedies for their patients—their main cognitive activity is rather one of diagnosis. However, their medical colleagues devising a new therapeutic protocol, for example, for treating cancer, may be considered involved in a design process. In the same way, a lawyer who is applying laws will not be considered a "designer," but the legal specialists drawing up new laws will.

The viewpoint according to which other people than so-called "designers" may be involved in activities that, from a cognitive viewpoint, are analyzed as "design," has another side. Indeed, a professional who is considered a "designer"—by management and colleagues—is not constantly involved in what we consider "design."

Without basing their critique on a cognitive analysis of people's activity, other authors have also advanced the idea that the term "designer" is sometimes used inappropriately. C. M. Burns and Vicente (1995), for example, judge that the term is applied "too narrowly" because the qualification ignores certain participants in the design project who "may be affecting the course of the design as much as any of the designated 'designers'" (p. 102). The prototype example of such an ignored participant is the user: This participant is considered increasingly a "designer" in approaches to design such as participatory design.

Schön also proposed to broaden the notion of "design professions," showing how professionals of many disciplines create what he called "design worlds" in the spirit of Nelson Goodman's *Ways of Worldmaking* (Schön, 1992, p. 4). Design worlds are "environments entered into and inhabited by designers when designing" (Schön, 1988, p. 181). The "indeterminate zones of practice" in which many professional practitioners are working, "especially . . . situations of uncertainty, uniqueness, and conflict," may be qualified as "design-like" (Schön, 1988, p. 181; see also Bucciarelli, 2002).

Without adopting a cognitive viewpoint, many designers themselves have a related viewpoint on this question. Many software designers want "software designer" to be an independent profession, recognized through a job title. They are concerned by the cleavage between their professional status as "software engineers" or "programmers," and their executing an activity that they consider "design."

In January 1991, the 15th-anniversary issue of *Dr. Dobb's Journal* (*DDJ*) focused on the design of software, with, among other papers, a manifesto by Kapor (reprinted as Kapor, 1996). This *DDJ* issue, especially Kapor's contribution, advocated that software design ought to be a separate discipline in the software domain. Kapor defends the specific role of design, drawing a parallel between fabricating software and fabricating buildings. According to the author, what we have "nowadays" (i.e., in 1991) are construction workers designing software whereas we need software architects. Such software designers need formal training and recognition as members of a separate profession, equal to computer scientists and engineers.

In 1992, a new professional organization was founded, the Association for Software Design (ASD). The ASD "delights in revealing to prospective members that they have been engaged in software design, even though their payroll records may refer to them as software engineers, as programmers, as program managers, as human-factors consultants, or as one of many other titles" (Winograd, 1996, p. xv).

In 1996, in his *DDJ* column "Design: Whose Job Is It, Anyway?," Swaine writes that over the last 10 years, in some organizations, "software designer" has become "a real job title," and that "there are schools offering multicourse programs in software design." Swaine notes that "books on software design aren't as common as books on Java programming," but there are several, "excellent" ones.

Finally, in 1997, Winograd mentions that, in the new field of "interaction design," there is the "new profession" of "interaction designer."

21.4.   THE OPPORTUNISTIC ORGANIZATION OF DESIGN: DECOMPOSITION AND PLANNING

We drew attention to the intermingled character of design and its surrounding tasks, namely requirements development and implementation. This interwoven character of activities also holds at a lower level, that is, inside the global design task. It holds with respect to both stages and abstraction levels.

The process followed by designers in industrial, complex projects does not progress through independent consecutive stages. Designers do not first structure the problem and then solve it: Design is not a process going from "analyzing the problem requirements" to "synthesizing the solution" (the strict SIP viewpoint). Neither do designers traverse systematically the three stages often distinguished in problem solving, namely construction of a problem representation, solution generation, and solution evaluation (a slightly different way of translating the SIP viewpoint).

With respect to progressively traversing abstraction levels, we concur with a conclusion formulated back in 1980 by T. R. G. Green in a paper on "planning a program." T. R. G. Green completes a discussion of structured-programming methods, stating:

> Good programmers.... leap intuitively ahead, from stepping stone to stepping stone, following a vision of the final program; and then they solidify, check, and construct a proper path. That proper path from first move to last move, in the correct order, is the program, their equivalent of the formal proof. (p. 306)

T. R. G. Green notes that the author who introduced the concept of "stepwise refinement" for program development, that is, Wirth, is himself "quite explicit." Having described this systematic decomposition strategy, Wirth says: "I should like to stress that we should not be led to infer that the actual program development proceeds in such a well organised, straightforward, top-down manner. Later refinement steps may show that earlier ones are inappropriate and must be reconsidered" (quoted in T. R. G. Green, 1980, p. 306).

This twofold conclusion—designing is systematic, neither in the traversal of problem-solving stages nor in the decomposition of problems—has been formulated, since the 1980s, in an increasing number of empirical design studies (Bisseret et al., 1988; Eckersley, 1988; Guindon et al., 1987; B. Hayes-Roth &

Hayes-Roth, 1979; Kant, 1985; Ullman et al., 1988; Visser, 1987b; Visser & Hoc, 1990). The designation that has come to qualify the way in which designers organize their activity "in the real world" is "opportunistic" (B. Hayes-Roth & Hayes-Roth, 1979; Visser, 1987a, 1987b, 1994a; Visser & Morais, 1991). This position, however, has also been the object of opposition (see the section *Discussion of Our Opportunistic-Organization Position*).

Like many colleagues (see e.g., Ball & Ormerod, 1995), we assume that designers have principles guiding their activity. These principles may have been learned, or result from standards imposed by their company or from personal preferences. Designers are aware that the use of a combined top-down breadth-first strategy is a valuable approach in order to organize their design properly: For example, to obtain well-structured specifications, think of all design components and handle correctly their interactions. They are conscious that avoidance of premature commitment is precious and that the breadth-first decomposition minimizes this risk. However, designers, for one thing, meet difficulties in the implementation of such systematic strategies and, for another, refer to other resources for organizing their activity besides the structured plans provided by systematic strategies. These strategies indeed impose a heavy load on memory and hinder possibilities for action that may be interesting for different reasons.

The absence of stages in design has been emphasized previously. In this section, designers' organization of their activity is examined in some detail. How do designers traverse the abstraction levels that *analytically* can be distinguished in the *result* of their activity?

In order to discuss these issues, we need to present design decomposition strategies and the corresponding planning of designers' activity. We therefore anticipate the general presentation of design strategies and first describe briefly the systematic decomposition approach advocated by design methodologies and two groups of strategies that may be used to accomplish such systematic decomposition (top-down and bottom-up, and breadth-first and depth-first strategies), and then discuss the different ways in which designers in fact plan and organize their activity.

Terminology. We distinguish between "plan" and "organization." "Plans" are mental representations that designers construct and use in order to anticipate, structure, and guide their design activity. The "organization" of designers' activity is their structuring their actual design activity.

"Plan" and "subplan" are relative notions (like "problem" and "subproblem"). When there is no risk of confusion, the term "plan" may refer to subplans (or plan components).

A Systematic Decomposition Approach to Design

Systematic problem decomposition is the overall strategy advocated by design methodologies for planning or organizing one's activity (these methodologies do not distinguish between the two). Different bases are proposed to perform such decomposition. In software design, the major paradigm is the integrated use of top-down and breadth-first decomposition strategies ("stepwise refinement," see later discussion).

Decomposition (or detailing, often qualified as "refinement") goes together with concretizing: Designers do not first decompose their problems until they obtain the most detailed subproblems, and only then attack concretization. The following presentation sometimes also mentions these concretization activities related to the implementation hierarchy, but we focus on decomposition.

By definition, designers will decompose the global design problem, made up by the artifact's requirements, in that they will convert it into several subproblems to be solved. This problem transformation can be analyzed as a traversal of the space constituted by an articulation of the two dimensions of abstraction (see Table 21.1).

The majority of classic empirical software-design studies present decomposition as an important strategy effectively adopted in design. As used in these early studies, "decomposition" generally conveys the presupposition that it takes place in a systematic way, proceeding from hierarchical planning and implementing balanced solution development. The traversal of the abstraction space corresponding to the problem transformation can indeed be reconstructed, retrospectively, as a systematic traversal that might have resulted from stepwise refinement (see Table 21.2).

Several analyses of these software-design studies have concluded, however, that their authors often confounded the structure of the *result* of the activity with that of the *actual activity* (Carroll & Rosson, 1985; Visser, 1987a; Visser & Hoc, 1990; Visser & Morais, 1991).

Top-Down and Bottom-Up Strategies

The top-down strategy—that is, its systematic implementation—consists of descending a problem's theoretical "solution tree" from the most abstract level down to the lowest, concrete level, never coming back up to a higher level. The

TABLE 21.1
The Two-Dimensional Abstraction Space to be Traversed From Start to Terminate

			Implementation hierarchy			
			Level 1	Level 2	Level 3	Level 4
			conc rete--	conc rete-	conc rete+	conc rete++
Part–	global+	level 1	*Start*			
whole	global	level 2				
hierar-	detailed	level 3				
chy	detailed+	level 4				*Terminate*

TABLE 21.2

Progression—From Start to Terminate Through the Two-Dimensional Abstraction
Space—That Results From a Systematic Implementation
of Combined Top-Down Breadth-First Problem Decomposition

			Implementation hierarchy			
			Level 1	Level 2	Level 3	Level 4
			conc rete--	conc rete-	conc rete+	conc rete++
Part–	global+	level 1	Start			
whole	global	level 2				
hierar-	detailed	level 3				
chy	detailed+	level 4				Terminate

bottom-up strategy consists of ascending the solution tree from the lowest to the most abstract level.

Design methods often propose that a systematic top-down strategy be applied iteratively. Early empirical studies conducted on design, especially in the domain of software design, claimed that designers indeed did so.

Also in an early empirical study on programming, however, R. E. Brooks (1977) predicted that a strictly top-down approach is only adopted by expert programmers who are familiar with the type of problem and have considerable experience with the programming language to be used.

It may be noticed that, like problem difficulty, familiarity and experience with problems and design tools are relative notions.

Breadth-First and Depth-First Strategies

In addition to top-down decomposition, design methodologies advocate that, in order to decompose a problem, one combine this strategy with a breadth-first strategy: When decomposing a problem solution, one should develop, consecutively, all the elements of the current problem solution at the same level of the solution tree (i.e., breadth-first) and then integrate the results into a new global structure. Following a depth-first strategy, one develops, consecutively, all the elements of one branch of the solution tree from its most abstract level down to its lowest, concrete level.

For handling interaction, breadth-first processing is of course very useful, even if the detection of potential interactions may require descending branches in anticipation.

Top-Down, Breadth-First Refinement

Several authors concluded that the approach advocated by design methodologies, that is, an integrated top-down, breadth-first decomposition, was generally the global strategy implemented by experts, whereas novices proceeded top-down, depth-first (experts sometimes did also) (Adelson, Littman, Ehrlich, Black, & Soloway, 1985; Adelson & Soloway, 1988; Ball & Ormerod, 1995; Byrne, 1977; Davies, 1991a; Jeffries et al., 1981).

However, this balanced, top-down refinement seldom occurred absolutely. Some 10 years ago, we analyzed the results of 15 empirical studies of design in diverse domains, focusing on designers' organization of their activity (Visser, 1992a, 1994a). Among these 15 studies, five concluded that designers implemented a *systematic* decomposition strategy—noticing just "a few exceptions" or "some deviations" by the programmers. Reanalyzing the results (based on the presentations by the authors), we concluded that, with one exception (Adelson & Soloway, 1988), each study showed one or more factors contributing to the opportunistic character of the organization of design. Jeffries et al. (1981), for example, who studied four experts, started their presentation of results asserting, "almost all subjects approached the problem with the same global control strategy," problem decomposition, by a progressive top-down expansion of the design, expanding it in a breadth-first manner at each successive level. A close reading of the paper showed, however, that only one expert implemented this "global control strategy" systematically; the other three deviated more or less from the strategy, both from its top-down and from its breadth-first component.

One may notice that data analysis (both of one's own data and of results published by colleagues) is an interpretational activity. When one analyzes data, one constructs representations of these data, and one thus focuses on, and selects, particular elements, guided by one's theoretical, sociocultural, and sociopolitical views.

For example, Ball and Ormerod (1995) and we (Visser, 1994a), each analyzing the Jeffries et al.'s (1981) software-design study, put different focus on the results presented by Jeffries et al. An example is the following. In our analysis, we emphasized that Jeffries et al. insisted on the systematic character of design, whereas we discovered several "exceptions" when closely reading the paper. Ball and Ormerod followed Jeffries et al., in first presenting, as a general outcome, the "highly systematic, top-down" design development, and only afterward noticing, "some expert subjects" "periodically violated" this approach.

With respect to designers' diverging from systematic decomposition, different sources of variation have been observed.

Handling a problem at one level, a designer may think of related elements at another level (cf. our factors of opportunism, presented later). Sometimes, experts are able to maintain this kind of elements in memory and retrieve them at the appropriate moment. Observing that the experts in their study make "notes to themselves" (concerning constraints, partial solutions, or potential inconsistencies), Adelson et al. (1985) posit the existence of "demons" reminding designers as to certain information that was skipped, to incorporate it into the design once the appropriate level of analysis has been reached.

These demons are not always effective, however. Several authors (Guindon et al., 1987; Hoc, 1988b) noted that even experts experience difficulties considering and maintaining simultaneously several problem-solution elements at one level of abstraction. They observed that subjects who were engaged in bottom-up processing activities encountered difficulties when backtracking of subproblems whose solution had been postponed or whose solution had to be modified.

Unbalanced development may also be the result of partial design solutions brought about, not by decomposition, but by memory retrieval as a result of recognition (Guindon et al., 1987; cf. our factors of opportunism, presented later). Both Darses (1994) and Visser (1990) also observed that, at repeated occasions during the design process, certain problem solutions already have a concrete, detailed form whereas others, which coexist with the previous ones, are still specified at a highly abstract, global level.

Hoc (1988b) evaluated a prototype programming environment that was to support top-down processing. He showed that professional programmers, trained in the underlying structured-programming method, had trouble due to the processing "imposed" by the environment, and generated nonoptimal solutions.

We conclude that empirical cognitive design studies show that *systematic* implementation of stepwise refinement is rare in practice. The traversal of the two-dimensional abstraction space corresponds rather to a movement as sketched in Table 21.3, than to the systematic progression drawn in Table 21.2.

Planning: Plan Elaboration and Plan Execution

"Planning" globally refers to two activities: Plan elaboration—that is, design of plans—and plan execution—that is, implementation of plans. These two types of activities are, of course, not independent. As holds for all design activities, plan elaboration will rarely proceed from scratch (Visser, 1994b). Plan execution often involves replanning—that is, modifying existing plan components, generating new components. Even if, out of context (before engaging in one's activity, on paper, or in one's mind) one may elaborate a plan, once "in action," this plan will at least evolve, if not change completely: One generally does not simply im-

TABLE 21.3
Imaginary, But Realistic, Example of a Traversal—From Start to Terminate—Through the Two-Dimensional Abstraction Space

			Implementation hierarchy			
			Level 1	Level 2	Level 3	Level 4
			conc rete--	conc rete-	conc rete+	conc rete++
Part–	global+	level 1	Start			
whole	global	level 2				
hierar-	detailed	level 3				
chy	detailed+	level 4				Terminate

plement as such the plans that one has elaborated. The remark, "design never ends"—neither with implementation, nor with maintenance, nor with actual use of the artifact—also holds for planning. People (not only designers) plan, and replan, and replan—and the actual organization of their activity has still a different structure than any of the underlying plans!

Most research on planning has been concerned with execution rather than elaboration of plans. The few studies that have been conducted on plan elaboration are now some 25 years old (Byrne, 1977; B. Hayes-Roth & Hayes-Roth, 1979; but see the more recent research presented in Chalmé et al., 2004) .

The design involved in plan elaboration differs, in several respects, from the design tasks generally analyzed in cognitive design research. Some of the contrasting dimensions are: The everyday versus professional character of the design task; its individual versus collective character; and the particular role of temporal constraints.

Compared to other design tasks, much planning is performed in nonprofessional, everyday contexts. Contrary to professional design tasks, it is often conducted individually. The studies on route planning (Chalmé et al., 2004; B. Hayes-Roth & Hayes-Roth, 1979) and on meal planning (Byrne, 1977) referred to earlier, examined such individual, free-time planning.

In addition to being characterized by the features just mentioned, route planning is often the affair not only of lay people, but also even of "novices," if not in the type of design per se (any driver may recurrently proceed to route planning), at least with respect to knowledge of the "application domain," that is, the environment to be traversed. It is exactly when they are planning to drive through an unfamiliar region that people need to plan their routes.

As all design, plan elaboration generally proceeds by both retrieval and construction. Plan retrieval may take the form of schema instantiation or reuse of plans, leading to retrieval of particular preestablished plans, which are memory representations that result from plan construction in previous activities. Plan construction proceeds by elaboration of new plans, also using preestablished plans that can be retrieved, in addition to other components.

Notice that, in a design context, "plan retrieval" (as opposed to plan construction) will rarely refer to retrieval of one global, complex preestablished structure that exists as such in memory. Such retrieval may occur in extremely simple tasks—that is, tasks that constitute a problem for nearly nobody. Design projects will, of course, comprise generally some of such simple tasks. These are not the design tasks, however, on which we focus in this book.

In addition, it seems useful to make explicit that, *even in simple tasks*, preestablished structures will only contribute relatively small components to more complex plan structures. Such complex structures will always need to be *constructed*. The complete plan corresponding to a complex activity can only be a reconstruction outside the activity situation: It will never correspond to a structure *that exists as such in memory*. We describe an example of such a complex plan.

Example. In our functional-specification study (see Table 6.1), we not only wanted to analyze the mechanical-design engineer's organization of his activity. We also wanted to compare this organization with the designer's plans—analogously to the comparison that ergonomists classically establish between people's task and their activity: One may indeed consider the designer's plans as his actual task. We therefore asked him to describe his activity. Confronting plans and organization has been the basis of our studies on opportunism (Visser, 1990, 1994a).

The mechanical-design engineer described his activity in the form of a hierarchically structured plan containing four levels (the global functioning of the automatic machine tool, its cycles, their components, and their descriptors) (Visser, 1990, Table 2, p. 259). According to this description, the engineer's control procedure for covering this tree structure reflected top-down planning, first breadth-first and then depth-first. Indeed, the engineer declared to first decompose the global functioning into the six different cycles implementing it, and to then further proceed depth-first for each individual cycle, specifying consecutively, for each component, all its descriptors.[30]

Hierarchically structured plans resulting from balanced, stepwise problem refinement may be interesting for reasons of a cognitive economy and because of the guidance they provide for a systematic activity. However, they are only one resource for designers among various action-proposing knowledge structures. If other structures propose actions that are more interesting, designers may deviate from their plans, or even not follow them. This is especially true for experts, who may be supposed to possess—or else to be able to construct rather easily—a representation of their activity that allows them to resume their plan later on, when, from the viewpoint of cognitive economy, it becomes more profitable to do so. Having several possibilities for action to compare and taking into account the cognitive cost of possible actions, are two task characteristics that probably only appear in real design. This may explain why in laboratory experiments mostly systematically organized design activities have been observed by our colleagues (Visser, 1990, 1994b).

Decomposition, Planning, and Organization

Even if they do no follow systematic decomposition strategies for planning their activity, designers usually decompose their problems in order to plan their activity. In general, the plan that results of such decomposition—be it a systematic or a nonsystematic decomposition—is, however, not systematically implemented in

---

[30] Visser (1990, p. 258) presented the engineer's control procedure for covering this tree-structure as reflecting "top-down, depth-first planning." This qualification *globally* translated the engineer's description of his functional-specification activity. To be completely *precise,* the described traversal started breadth-first, and then continued depth-first, as presented in this text. We add this details because our presentation of the designer's description as purely depth-first played a decisive role in Ball and Ormerod's (1995) discussion of our view on opportunism.

the design activity (see, e.g., our software-design study, Visser, 1987b, presented in the following example).

Example. The software engineer whom we observed in our software-design study (see Table 6.1) decomposed his activity according to two principles. For one thing, he divided his global IPC task into three consecutive parts according to the relative urgency with which various colleagues (especially in the work-shop) needed these different parts. For another, he decomposed the IPC software that was to be developed, into various modules, corresponding to different machine-tool functions. This breakdown followed the order in which modules appeared on the listing of software that the designer had previously developed for a similar machine-tool installation and that he reused here (the designer's "example" software). This decomposition guided his consecutive planning of each of the three parts resulting from his task planning based on urgency.

After this planning phase (which took him only 1 hour), the designer started to code. Further planning during this coding was local. It took place at various problem-solving levels and concerned more or less large entities (e.g., at the design level, it could concern a function or a machining operation; at the coding level, a module or a line of software). The engineer frequently deviated from the plan based on the example program.

Therefore, except for local planning occurring during coding, the only planning performed by the software engineer was guided by the urgency of components and by an example program listing, which the engineer followed for reuse. The organization of the designer's activity did not realize, however, this example program's plan.

Other authors have observed this absence of systematic decomposition and planning. Détienne (1994), for example, observes a succession of planning and coding processes at a micro level. If one would like to qualify these processes in terms of design "stages" or "phases," they make up "miniature" phases of planning and coding.

Such a scattered way of proceeding, giving activity an unstructured organization, is not due to designers being nonchalant or incompetent. "Disciplined" structured approaches, notes T. R. G. Green (1990), referring to the waterfall model, "seem to be unfeasible as a general technique, because the consequences of higher-level decisions cannot always be worked out fully until lower level ones are developed" (p. 119).

The Opportunistic Organization of Design

B. Hayes-Roth and Hayes-Roth's (1979) study on errand planning is generally *the* reference for the view on design that qualifies its organization as opportunistic. The authors described their subjects' "largely opportunistic" planning activity in the following terms:

At each point in the process, the planner's current decisions and observations suggest various opportunities for plan development. The planner's subsequent decisions follow up on selected opportunities. Sometimes, these decision-sequences follow an orderly path and produce a neat top-down expansion. . . . However, some decisions and observations might also suggest less orderly opportunities for plan development. . . . Interim decisions can lead to subsequent decisions at arbitrary points in the planning space. (p. 276)

This description sounds informal. "Opportunism" is, however, not just a "blanket-term to denote *any* design activity that deviates from a single, rigid design approach" (p. 137), as Ball and Ormerod (1995) consider opportunism proponents to do. B. Hayes-Roth and Hayes-Roth (1979) modeled the activity using a blackboard model, that is, a special case of opportunistic problem-solving models (Nii, 1986a, 1986b; see also B. Hayes-Roth et al., 1979). In the authors' model, a tentative solution, that is, a tentative plan (i.e., a provisional design solution) is elaborated by several cooperating cognitive "specialists" (knowledge sources) making decisions concerning which they can interact and communicate via the "blackboard." This common data structure is partitioned into several planes containing conceptually different categories of decisions ("planes"), that is, for the route planning analyzed by B. Hayes-Roth and Hayes-Roth, the plan, executive, meta-plan, plan-abstraction, and knowledge-base planes. The assumptions of the model are illustrated and corroborated with a subject's verbal protocol. The subject's processing was, indeed, multidirectional: The sequences of his basic actions, that is, his decisions, included both top-down and bottom-up instances; plans, in addition to being formulated at abstract, high levels, were also formed at low levels in the absence of corresponding higher level plans (p. 306).

We expanded this view of opportunism for more "typical" design tasks (functional specification and software design, see Table 6.1; see also Bisseret et al., 1988, for traffic-signal setting; Whitefield, 1986, for mechanical design). We have focused on designers' organization of their activity, analyzing its underlying cognitive foundation (Visser, 1988c, 1990, 1994a).

We attribute the particular nonsystematic character that we qualify as opportunistic to the fact that designers, rather than systematically implementing a structured decomposition strategy, take into consideration the data that they have at the time: Specifically, their knowledge, the state of their design in progress, their representation of this design, and the information at their disposal (Visser, 1990, p. 267). Considering this data in addition to the possibilities provided by systematic decomposition, designers considerably increase their potential range of action.

It may be important to emphasize that—even if we sometimes use the abridged expression "design is opportunistic"—the notion "opportunistic" in our work applies to the *organization* of designing, not to individual design activities (Visser, 1990, p. 267). Opportunism is not a strategy among others. In terms of "strategies," opportunism might be qualified a "meta-strategy," but we consider it more appropriate, because it is more explicit, to adopt the notion "organization."

Our Analysis
of the Opportunistic Organization of Design

In this book, we briefly introduce the organizational component of the global model for which we are aiming (for more details, see Visser, 1988c, 1990, 1994a).

Globally qualified, we distinguish an action-execution (AE) level and an action-management (AM) or control level, structured according to the following iterative sequence:

- AM: Call for action proposals.
- AE: Proposal of one or more actions.
- AM: Selection of one action, based on evaluation of proposals.
- AE: Execution of the selected action.
- Go back to AM: Call for action proposals.

As regards action proposal, opportunities must be identified as such. Their perception as opportunities is concept-driven and data-driven. It is indeed based on their knowledge and the representations they construct, that expert designers process the data that they perceive in their environment and that may take different forms (see the six factors of opportunism presented later). Taking advantage of these opportunities rather than following a preestablished plan resulting from a systematic decomposition strategy will lead to an opportunistically organized activity.

As regards action selection, we have identified "cognitive economy" as the central cognitive evaluation criterion of opportunities' "interestingness." Indeed, potential "opportunities" are only exploited effectively if they are interesting from a cognitive viewpoint, that is, if they are cost-effective (viewpoints other than cognitive have not been analyzed in our studies of design organization). This evaluation of a proposal's interestingness is of course a relative assessment that compares not only the competing proposals to each other, but also takes into account possible actions envisioned more ahead in the process.

In our studies, we identified six types of data (information and mental representations) that could be "taken advantage of" as factors leading to the opportunistic organization of design (Visser, 1990, 1994a):

- Information provided to the designer by an external information source (in particular, the client or a colleague).
- Information the designer "comes across" when "drifting" (i.e., involuntary attention switching to a design object or action other than the current one).
- Mental representations constructed by the designer when considering current data from another viewpoint.

- Mental representations constructed by the designer as by-product of the current design action.
- Mental representations of design objects activated by the representations used for the current design action, because of activation-guiding relationships existing between the two sets of representations.
- Mental representations of design procedures activated by the representations of the procedures used for the current design action.

Guindon et al. (1987) qualify the design activity that they analyze as "serendipitous," observing that problem solving is controlled by recognition of partial solutions, at different abstraction levels of the solution, without designers having previously decomposed the problem into subproblems. This case of serendipitous design may be occasioned by our fifth factor of opportunism.

We present next an example from our functional-specification study (see Table 6.1).

Example. This example illustrates designers taking advantage of mental representations of design objects connected by the relationship of analogy to the representations that they are using for their current design action (our fifth factor).

The designer had planned to specify first-phase tooling operations (used in order to shape the rods) before second-phase tooling operations (used in order to finish the rods). The actual specifications of the first-phase and that of the second-phase operations are completely intertwined, however. Considering second-phase operations as analogous to first-phase operations, the designer, during his working on the first-phase tooling function, continually switches between first- and second-phase specifications. Often he takes advantage of the specification of a first-phase operation O1 in order to specify, by adaptation of this O1 specification, its corresponding second-phase operation O2. Frequently, an O2 specification "makes him think" of an omission or error made on the corresponding O1 operation.

The observation that designers organize their activity in an opportunistic way is not restricted to inexperienced designers. On the contrary, we concluded that it is something typical of expert designers (Visser, 1992a). Nor is it the translation of a deteriorated behavior that occurs only when designers are confronted with a "difficult" design task. Even when expert designers are involved in routine tasks, the organization of their actual activity is not well-structured (Visser, 1994a).

Discussion of Our Opportunistic-Organization Position

Until now, we have presented the opportunistic organization of design as *the* conclusion that is imperative if one analyzes critically the empirical data available since the early years of cognitive design research. There are, however, researchers who do not share this position and defend alternative views (especially, Ball & Ormerod, 1995; Davies, 1991a).

We assume, however, as do Ball and Ormerod (1995), for example, that top-down pursuit of design goals is a basic guiding principle for designers' activity (Visser, 1990). Yet, designers may deviate from such a structured way of proceeding. We observed them to do so, and we identified and analyzed the factors underlying these deviations and the resulting organization of their activity!

Several researchers have presented nuanced positions with respect to design organization. Some authors have observed that designers' activity may be organized hierarchically if the designers possess the corresponding schema, but that otherwise their activity is opportunistically organized (Rist, 1990). Some did not use the term "opportunistic," but noticed that they observed "no evidence of a well-developed solution plan" and that designers did not proceed to systematic decomposition, but "encountered" subproblems when "exploring the implications of a proposed solution" (Voss et al., 1983). Sometimes, participants in experiments "claimed to have designed top-down," but the researchers did find no correlation between the "self-reported strategy variables" and the final designs (Malhotra et al., 1980) (cf. also the difference that we noticed between the plan described by the designer as the structure of his activity, and the organization of his activity observed by the researcher, Visser, 1990, 1994a). Goel (1995) presents design as a quite systematic process, but he also remarks that "designers differ substantially in the path they take through [the design problem] space and how quickly or slowly they traverse its various phases" (p. 123). In addition, he notices that "problem structuring" (in his model, the first phase of design development) "occurs at the beginning of the task, . . . but may also recur periodically as needed" (p. 114).

Davies (1991a) presented his position in a paper titled "Characterizing the Program Design Activity: Neither Strictly Top-Down nor Globally Opportunistic"—the opposite of what we proposed, qualifying the "Organization of Design Activities [as] Opportunistic, With Hierarchical Episodes" (title of Visser, 1994a). Davies states "novice programming behaviour appears to be *systematically opportunistic*, displaying none of the characteristics of top-down design. Conversely, expert programmers adopt a *broadly top-down* approach to the programming task, *at least during its initial stages*" (p. 186; the emphasis is ours). Opportunistic episodes appearing as the task progresses was also observed by Ullman et al. (1988).

As regards the "broadly top-down with opportunistic local episodes" or "opportunistic, with hierarchical episodes" issue, we follow B. Hayes-Roth and Hayes-Roth (1979, p. 307). In order to solve "the apparent conflict" between the top-down, successive refinement model and the opportunistic model, these authors have proposed that the refinement model be considered as a special case of the opportunistic model. Indeed, a model that allows various organizational structures of an activity (an opportunistic model) is more general than a model that allows only one, or a combination of two structures (top-down combined with breadth-first or depth-first). An opportunistically organized activity may have hierarchical episodes at a local level, but its global organization is not hierarchical (Visser, 1994a).

With respect to these modeling choices, several questions may be raised. How great a lack of systematicity can be tolerated by a systematic-decomposition model? Alternatively, what degree of systematicity can be qualified as a particular form of "opportunism" before this qualification becomes inappropriate (even if formally, one can always model it as an opportunistically organized activity)?

Yet, we favor an opportunistic model not only because of this formal argument, but also for "positive" reasons (see later discussion).

In his conclusion, Davies (1991a) notes that the differences between experts and novices may be due to differential "use and reliance upon external memory sources" (p. 186). This point becomes central in Ball and Ormerod's (1995) discussion of design organization, focusing on research by Guindon (1990a) and Visser (1990) as authors defending the opportunistic stance. Ball and Ormerod argue, "the existence of opportunism in expert design has been exaggerated" (p. 131). They claim, "much of what has been described as opportunistic design behavior appears to reflect a mix of breadth-first and depth-first modes of solution development" (p. 131), even if design is also "subject to potentially diverging influences such as serendipitous events and design failures" (p. 145).

Our view on these positions is that we do not deny that designers may proceed in a breadth-first or depth-first way, but we want to emphasize that

- they often do so occasionally and locally, rather than systematically and throughout their design process.
- a breadth-first - depth-first mix can take different forms: Even if a mix pattern has several occurrences, these will generally be interspersed with other ways of proceeding, involving the risk that "breadth-first" and "depth-first" are no longer applicable as general qualifications of a designer's activity or strategy; and especially
- an occasional, local breadth-first - depth-first mix is just one form in which opportunism can reveal itself in design.

If design was in effect "simply" "a mix of breadth-first and depth-first modes," it might indeed be exaggerated to resort to opportunism, which in that case would open up a needless large space of possible forms of structuredness and unstructuredness.

Example. In our functional-specification study, there was one series of breadth-first episodes occurring several times consecutively. These episodes consisted of switches between two workstations, WS1 and WS2, in order to define tooling operations on WS2 by analogy with those on WS1. The designer indeed considered these two workstations as analogous (cf. the example presented in the subsection *Our Analysis of the Opportunistic Organization of De-*

sign). This was the only systematic in-breadth movement during the entire functional-specification design process.

With respect to the structured character of design organization, opportunism proponents (Guindon et al., 1987; Kant, 1985; Ullman et al., 1988; Visser, 1987a; Voss et al., 1983) question not only the systematic implementation of breadth-first and depth-first approaches, but also of systematic top-down refinement, an issue that Ball and Ormerod (1995) omit completely from their discussion. However, Jeffries et al.'s (1981) results, which are one of the foundations of Ball and Ormerod's (1995) claim, show that the top-down strategy also is not systematically implemented by expert designers.

We close this section by a short discussion of the "potential causes of unstructured activity" (Ball & Ormerod, 1995, p. 145). Ball and Ormerod concede that "like Guindon (1990) and Visser (1990) [they] recognize a number of [such causes]. These include factors such as social influences, memory failures, design failures, information unavailability, boredom and serendipitous events" (p. 145).

We suppose that most, if not all, cognitive functioning is, for an important part, under the dependence of people's limited information processing capacities (cf. Simon's bounded rationality). We suppose, however, that in addition to the more or less frequently arising *impossibility* of working in a systematic way because of cognitive *deficiencies*, there are also positive causes of opportunism. Designers also *decide* to take *advantage* of opportunities, in spite of, or by preference to other possibilities. The decision to exploit analogical relations, information available because a colleague passes by, or "discoveries" that emerge as by-products of systematic design actions, are all examples of instances of such positive factors leading to an opportunistically organized activity.

21.5.   DESIGN STRATEGIES

Given the nature of this book, we focus on strategies that are used by experienced, professional designers.

During the second half of the 1980s, several studies in the domain of "psychology of programming" have examined design strategies, especially the systematic decomposition strategies discussed previously. These studies concerned, however, mostly students, that is, novices. Moreover, even when they studied "experts," these were often advanced students rather than professionals.

There are, however, several studies identifying strategies used by expert, professional, designers (Ö. Akin, 1979/1984; Gilmore, 1990b; Visser & Hoc, 1990).

The following subsections present some data on design strategies used by professional designers, namely reuse, simulation, selecting a kernel idea and premature commitment to the idea, and user considerations guiding development and evaluation. For other strategies (e.g., data-driven and goal-driven, prospective and retrospective, and procedural and declarative strategies), see, for example, Visser and Hoc (1990).

Reuse Versus Design "From Scratch"

All use of knowledge could be qualified "reuse" in that knowledge is based on the processing of previous experience and data encountered in the past. We have proposed to reserve "reuse" (vs. other "use" of knowledge) for the use of specific knowledge that is at the same abstraction level as the "target" for whose processing this knowledge is retrieved (Visser, 1995a). "Reuse" of knowledge is thus opposed to the use of more general, abstract knowledge (such as first principles and knowledge structures like schemas and rules).

Terminology. In the research literature on analogical reasoning and reuse, "target" is the term adopted for the problem to be solved (or, in other than problem-solving tasks, the critical material to be processed). "Source" is the term adopted for the knowledge used in order to solve the target problem: In analogical reasoning and reuse, it is generally a solution to another, analogous problem (a "case" in terms of case-based reasoning, CBR). Candidates for sources are often supposed to be solutions to problems that are similar ("analogous") to the target problem. The main criterion is, however, their usefulness for solving the target problem.

Reuse has been identified in various empirical design studies. The exploitation of specific experiences from the past is indeed particularly useful in design. Even if this may seem surprising, reuse is especially valuable in nonroutine design (Visser, 1995a, 1995b, 1996; Visser & Trousse, 1993; and, hereunder, the subsection *Reuse, Creativity, and Novelty*).

In AI, a phenomenon similar to reuse has been analyzed in terms of CBR. The CBR community also has argued that this form of reasoning is particularly well suited to design (see, e.g., Pu, 1993) (for a cognitive-psychology analysis of these issues, see Visser, 1993c, 1994d; Visser & Trousse, 1993; see also A. Heylighen & Verstijnen, 2003) (for differences between analogical reasoning, reuse, and CBR, see Visser, 1999).

Before cognitive-psychology and cognitive-ergonomics researchers started to examine reuse, their colleagues in computer science had already identified and examined this research topic. Authors in the software engineering community had defined the domain of "software reuse" (Biggerstaff & Perlis, 1989b, 1989c).

Software-engineering researchers distinguish "design for reuse" from "reuse for design" (Thunem & Sindre, 1992). The construction of reusable entities that are to be organized into a "components library" (design for reuse) is often considered an independent design task, not necessarily executed by the designer who is going to reuse these entities (reuse for design). We are unaware, however, of any empirical study conducted specifically in such design-for-reuse situations. Existing empirical studies concern reuse for design, and show that the two activities are not as separate as software-engineering researchers may suppose.

A considerable proportion of the empirical, cognitive-ergonomics research on reuse in design has been conducted in the domain of software, especially that of object-oriented software (Burkhardt, 1997; Détienne, 2002a), but other paradigms have also been examined. We have observed reuse in IPC-software design by a designer using a declarative type of Boolean language (Visser, 1987b, 1988b; see Table 6.1). Weber (1991) has examined reuse in programming with LISP. We have also studied reuse in other domains of design, namely mechanical (Visser, 1991b) and industrial design (Visser, 1995b).

Several aspects of reuse have been examined in these studies, such as its phases and strategies, types of entity reused, types of exploitation of reusable entities, effects of reuse on designers' productivity, and difficulties and risks of reuse (see, e.g., Détienne, 2002a; Visser, 1995a). Here, we focus on the question that is at the basis of reuse-based design: When do designers adopt reuse in order to solve a design problem, rather than base their activity on general knowledge, that is, "design from scratch" (or "depart from a *tabula rasa*"; Tzonis, 1992)?

Reuse takes place in, at least, five phases:

- Phase 0. Construction of a representation of the target problem.
- Phase 1. Retrieval of one or more sources.
- Phase 2. Adaptation of the source material into a target-solution proposal.
- Phase 3. Evaluation of the target-solution proposal.
- Phase 4. Integration into memory of the resulting modifications in problem and solution representations.

One may notice that this five-phase model is a particularization of the general three-phase problem-solving model—with a memory-integration phase appended:[31]

- Phase 1. Construction of a problem representation (cf. reuse phase 0).
- Phase 2. Solution generation (cf. reuse phases 1 and 2).
- Phase 3. Solution evaluation (cf. reuse phase 3).

This comparison shows the attention paid in research on reuse to the solution-generation mode that is particular to reuse, that is, the solution-generation mode used in analogical reasoning.

The construction of a target-problem representation (phase 0) has seldom received any attention in empirical studies, neither in studies on analogical reason-

---

[31] Such a memory-integration phase normally receives not much attention in problem-solving research. This may be a consequence of "memory integration" depending on the "learning" rather than the "problem-solving" department in psychology. Classically, one distinguishes "functional departments" in psychology, such as memory, language, problem solving, and learning (or development).

ing nor in those on reuse. It is, however, during this phase that designers have to decide whether they are going to try reuse in order to solve their design problem. We are not aware of empirical studies that observed if designers, rather than to "simply" design from scratch right from the start, consider the choice between design from scratch and design based on reuse. Two studies seem to indicate that the cost of reuse is an important, if not the main factor in this decision process (Burkhardt, 1997; Visser, 1987b).

In an experimental study, Burkhardt (1997) asked seven object-oriented (OO) software designers to describe elements that they might want to use or reuse. Half of the participants mentioned that there were reus*able* elements whose actual reuse they would not envision because of the cost of their reuse.

Data that we gathered on reuse in software design (Visser, 1987b) provides a factor contributing to the cost of reuse—but only once have candidate sources (reusable solutions) been retrieved. This factor is the cost of required adaptation, itself a function of target-source similarity. Sources are indeed always to be adapted in order to be usable as a possible solution to a target problem. This conclusion coincides with a position adopted for several CBR systems in which the selection of a case, that is, a reusable source, is guided by its adaptability (Keane, 1994; Lieber & Napoli, 1997; Smyth & Keane, 1995).

The importance of the cost factor in the choice of a strategy such as reuse is in line with our identification of the primordial factor underlying the organization of design, namely cognitive economy (Visser, 1994a). If designers are conscious of several action possibilities, the relative cost of an action is a decisive factor in their choice.

With respect to the frequency of reuse actually taking place, different authors in the domain of software engineering assert that between 40% and 80% of code is nonspecific, thus reus*able*. Many authors advance the percentage of 80%, mostly referring to an old paper by T. C. Jones (1984), who summarizes four studies conducted between 1977 and 1983. In 1989, however, Biggerstaff and Perlis asserted that, in fact, "over the broad span of systems, reuse is exploited today [that is, in 1989] but to a very limited extent" (Biggerstaff & Perlis, 1989a, p. xvii).

Empirical studies providing quantitative data about actual reuse all concern OO software, which is considered to particularly "favor reuse," due to its mechanisms of inheritance, abstraction and encapsulation, and polymorphism. The conclusions of these studies may thus be specific to this particular software-design paradigm. As far as we know, the only empirical study observing "massive" reuse indeed concerns a field study on OO software by Lange and Moher (1989). No other empirical design studies provide us with data on the frequency of reuse: There are thus no studies at all in other domains than OO software design concerning this issue.

Studies on analogical reasoning might, however, be informative—even if they have been conducted in the laboratory, thus in artificially restricted conditions. The general conclusion of these studies is that source retrieval seldom occurs "spontaneously," that is, without being suggested from outside. In laboratory studies, the experimenter often carries out this clue function. In professional work situations, however, one cannot expect that external persons urge designers to reuse — a possibility is that reuse is an enterprise-policy issue.

The data available on reuse-based design versus design from scratch is thus still very meager. Nevertheless, it is a central topic with respect to reuse, its role in design, and the possibility to support designers in their reuse during design (Falzon & Visser, 1989).

Reuse, creativity, and novelty. In the subsection discussing the role that Simon attributes to decomposition (*Overestimating the Role of Systematic Problem Decomposition*), we remarked that systematic decomposition generally leads to standard, routine solutions because of its top-down nature. This remark should not lead one to believe, however, that bottom-up or data-driven problem solving is a guarantee of novelty, whereas a top-down or concept-driven way of proceeding necessarily leads to routine solutions.

Example. In an analysis of data collected in several of our studies, we have identified two different types of analogy use (Visser, 1996). One of our conclusions of this analysis was that, according to the use that is made of analogical reasoning, this form of reasoning may, or may not, add to the innovative nature of the activity (creativity) and of its result (novelty). Augmentation of novelty generally seems to result from analogical reasoning at the action-execution level, that is, the type of analogy use traditionally examined in cognitive-psychology studies. The second eventuality occurs with analogical reasoning at the action-management level. Especially on the level of structure, the transfer of previously applied solutions into a new design project may introduce homogeneity into that project and, because of that, remove a possibility of novelty in the resulting design (see also Edmonds, Riecken, Satherley, Stennin, & Visser, 1994).

In their study of opportunistic planning, B. Hayes-Roth and Hayes-Roth (1979) observe that, compared with a fixed, one-directional planning approach, "the bottom-up component in multi-directional processing provides a potentially important source of innovation in planning" (p. 306). The authors also refer to Feitelson and Stefik (1977) who made similar remarks concerning the "largely event driven nature of the planning process" used by an expert geneticist whom they observed. The expert seemed to proceed in this event-driven way because of his "fishing for interesting possibilities" (Feitelson & Stefik, 1977). Based on our previous observations, we wish to emphasize that data-driven problem solving indeed constitutes a "*potentially* important source of innovation," but that one should not conclude that it is always, or by definition, leading to more innovation than a concept-driven way of proceeding!

Tzonis (1992) has proposed what he qualifies as a "precedent-based" design approach, working from the example of analogies used by Le Corbusier. This famous architect exploited analogies with huts, ships, and bottleracks, in order to design his famous *Unités d'Habitation*. Tzonis presents this precedent-based approach as an alternative to the "analytical paradigm" such as it had been elaborated in the early days of the computer introduction in architecture, initiated, in the 1960s, by architects such as Christopher Alexander (see earlier discussion) and Serge Chermayeff (with whom Tzonis himself had collaborated during those years). It is not surprising that the characteristics attributed by Tzonis to the analytical paradigm parallel those that we consider as characterizing the SIP paradigm. Neither is it amazing that Tzonis considers that analytical techniques may be applied to well-defined problems, but are of no use for problems that ask for creativity.

Tzonis underlines the important role of domain knowledge in such an approach—and thus the challenge and difficulty of gathering enough knowledge!

In the special issue of *Design Studies* on visual design representation, mainly focusing on sketches and other types of drawings, Eckert and Stacey (2000) presented various types of visual representations as "sources of inspiration." In a study on knitwear design, Eckert and Stacey examined the reuse of previous knitwear designs, but also of other objects and images that functioned as sources for design ideas: These were textiles and other decorative products, works of art, and objects and phenomena from nature and everyday life (such as flowers, leaves, or tiles). Designers explored these sources for shapes, patterns, motifs, and color combinations.

The authors analyzed these sources as elements of a "language of design" with its syntax[32] and its semantics. Knitwear designers used this language not only as a thinking tool for themselves, but also as a means to communicate their ideas to others. The authors note that such communication of design ideas by reference to designers' sources of inspiration plays not only an important role in this aesthetic form of design that is knitwear design, but also in technical domains, such as helicopter design.

When designers use such sources, they specify particular, individual designs as combinations, modifications, and adaptations of one or more sources. Interpreting such references requires a redesign activity on behalf of the designers' interlocutor—who is thus supposed to share the designers' sources and their cultural context. This is not the case of each interlocutor: In fact, only fellow specialists will adequately understand all references, but superiors and customers may have difficulty following designers' explanations based on sources of inspiration. Complementary colleagues, such as the technicians who are to manufacture the garments, may also lack essential references, so that they often interpret designers' specifications in terms of the company's designs of previous seasons. This may contribute to explain manufactured garments being "more conservative than their designers intended" (Eckert & Stacey, 2000, p. 535). Even more in general—and that is one of the authors' conclusions—this "language of design biases new designs towards existing ones, because the more a new design differs

---

[32] The authors describe this syntax in terms of *nouns* and *adverbs*, corresponding respectively to the sources and their modifications. They observe that the knitwear designers' source language does not have *verbs*. Analyzing the language in terms of a *predicate(arguments)* structure might have avoided this deficiency.

from the stock of old designs, the harder it is to imagine it or express it" (Eckert & Stacey, 2000, p. 538).

Simulation

Simulation is often presented as an important evaluation strategy. For Adelson et al. (1985), who work in the domain of software design, mental simulation of a design is the main operator used to check the sufficiency and the consistency of the current design state. In order to evaluate tentative solutions, designers run simulations and select between them, based on the criterion of efficiency, for example.

Simulation is also used in solution generation. Mental simulation may serve, for example, to understand and elaborate the problem requirements (Guindon et al., 1987; Kant, 1985; Ullman et al., 1988) or to access in memory relevant knowledge concerning a problem (Baykan, 1996, p. 141). Adelson et al. (1985) also observe that software designers use simulation to expand their design from one abstraction level to the next. The notion of "mental model" is often advanced in this context as the cognitive device that allows this processing. Adelson et al. observed that their subjects elaborated mental models at different levels of abstraction, and then run them at these different levels. These simulation runs were supposed to assist the designers on, at least, two points:

- Predicting potential interactions between elements of the design.
- Pointing out elements of the solution that needed expansion.

One may distinguish, at least, two types of mental simulation: Time-driven and event-driven processes. In addition, simulation may proceed at different levels of abstraction (B. Hayes-Roth & Hayes-Roth, 1979, p. 284).

In design domains involving physical artifacts (e.g., architectural, mechanical, and industrial design), designers, in addition to mental simulation, also proceed to simulation using physical devices, that is, external representations such as drawings and mock-ups.

In a comparative analysis of manual and computer-assisted design, Dillon and Sweeney (1988) observed the utility of such physical simulation—but only for some of the manually working designers. "Movement of parts of the design was often simulated on the board by using cut-outs of the relevant shape and literally moving it around the drawing by hand. This acted as a powerful visual aid to some designers. No directly comparable facility existed on CAD" (p. 484).

Simulation of a solution also plays a particularly important role in design because of the impossibility to test the artifact specified by the solution. An artifact product can only be tested by way of "testing" its representations, that is, the artifact product *as envisioned*. These representations can take a physical form (e.g., prototypes and mock-ups)—but they are not the artifact products. Both when such representations are mental and when they are physical, testing will be realized through simulation.

Selection of a Kernel Idea
and Premature Commitment to the Idea

Many authors have observed that designers, early on in their activity, tend to select a "kernel idea" (Guindon et al., 1987; Kant, 1985; Rowe, 1987; Ullman et al., 1988) and stick to it in what is going to become their global design solution. "Primary generator" is the expression proposed in 1979 by Darke, the first author to describe that designers, at the very start of a project, adopt a few simple objectives in order to generate an initial solution kernel.

Ullman et al. (1988) studied how different mechanical-design problems were solved by experienced mechanical designers who could be expected to have a high degree of expertise for the selected problems. The authors observe that the designers, rather than to start by formulating a global design plan, develop and gradually extend a "central concept" (p. 36).

Kant (1985) makes a similar observation, noting the importance of the "kernel idea": Design generally starts by focusing on an idea that is "quickly selected from those known to the designer . . . [who] lays out the basic steps of the chosen idea and follows through with it unless the approach proves completely unfeasible" (p. 1362).

Guindon et al. (1987, p. 64) observe that designers in their study focused quite early on in the design process on a kernel solution based on a primary generator that directed the complete design process and all the generated partial solutions.

Many different notions are more or less related to the kernel idea. In addition to "primary generator" and "central concept," other notions are "early solution conjectures" (Cross, 2001a) and "primary position," that is, "a small set of criteria which guides the adoption of an initial solution and provides guiding principles in developing the solution." One also finds references to "position-driven" design as designers establishing some critical issues and using this "position" to select solutions for many of the subproblems identified. In architecture, one speaks of a "concept" that underlies a project and is critical in the reception of the project (Lawson, 1994).

The process of selecting early on in the design process a kernel idea and then sticking to it has been qualified as a form of "early fixation," "premature commitment," "early crystallization" (Goel, 1995), or "solution fixation" in "problem formulation" (Cross, 2001a). There are studies on "design fixation" conducted from a slightly different perspective. In this context, "fixation" refers to designers who, when provided with pictorial representations of possible sources for the envisioned artifact product, tend to become fixated on these sources (Purcell & Gero, 1996).

Given the diverse observations on selection of a kernel idea, one might be tempted to conclude that designers generally neglect to elaborate and confront alternative representations of design problems and solutions. There are, however, different accounts of this issue. Several authors notice that designers often try to avoid early fixation (Lebahar, 1983; Reitman, 1964, p. 293).

In a study on professional design review meetings, Ball and Ormerod (2000) conjecture that it is the individually conducted or de-contextualized character of design that induces early fixation on a kernel idea. In their earlier studies of individual designers, Ball and Ormerod observed premature commitment comparable to the results presented earlier in this section. In these studies, they found "evidence for satisficing in the selection and evaluation of putative solution options. Individual designers frequently [became] fixated upon single solution ideas (usually derived from prior experience) rather than exploring alternatives in order to optimise choices . . ., even when they [recognized] that the solutions [were] less than satisfactory" (p. 157). In the authors' study of design review meetings, however, they observe how "a design team [explores and critically appraises] a range of alternative design concepts in an in-depth manner" (p. 164). The results of this study are considered "surprising" by the authors "in the degree to which they contrast so markedly with [their] previous studies of individual and de-contextualised designers" (p. 163). In the design-review meetings study,

> almost all episodes appear to reflect a motivated attempt by designers to generate and evaluate multiple solutions options . . . with the aim of maximising the choice of a best alternative in relation to a particular design question. . . . This evidence for optimising attempts at multiple solution search and evaluation is a finding that generalises across the different companies [Ball and Ormerod] researched and the diverse design tasks that were being tackled within these companies. (p. 163)

It was the team manager who, in the Ball and Ormerod (2000) design-review meetings study, acted as a

> safeguard against premature commitment to single-solution options should either the project champion or the group of collaborating designers become overly fixated upon a particular idea or concept prior to a full and cautious exploration of viable alternatives. It appears that the team manager has a very clear and explicit notion of the importance of ensuring multiple solution search and evaluation such that they are seen to perform an important steering function when designers get fixated upon single or unsatisfactory solution ideas or need support in pursuing alternatives. (p. 164)

One may relate this result to observations we made in our study on industrial composite-structure design by a team of mechanical designers (Visser, 1993a, see Table 6.1). In this study, the two most experienced designers were observed to come up with a multitude of solution principles.

In a situation that may be considered "de-contextualized," R. P. Smith and Tjandra (1998, quoted in Cross, 2001a) have studied teams of designers working on artificial design exercises. They also observed some teams being willing to reconsider early concepts. The authors concluded that, with respect to the quality of design solutions produced, they seemed to be dependent on this designers' readiness (Cross, 2001a, p. 87).

On the other hand, there are also studies observing individually working designers to come up with several solution ideas. The industrial designer whose protocol is analyzed by Eastman (1969, experimental design of a bathroom) does not seem to stick to one "kernel idea." He generates and evaluates five alternative solutions—two of which are finally going to be completely developed. More in general, however, the designers examined by Eastman proceed right from the start by specifying design elements at a concrete, rather than abstract level.

In his analysis of fugue composition by an individually working composer in his natural working context, Reitman (1964) observes that alternate solutions and plans may coexist for a time, with the decision between them deferred, pending the outcome of work on the material that may serve as a subcomponent of one or the other solution, or perhaps both. The composer was sometimes observed to oscillate for a considerable time between several alternates (pp. 309–310).

In a study comparing two types of individually working designers, that is, CAD-using mechanical designers and manually drawing designers, Whitefield (1989) observes that the manually drawing designers take into consideration more solutions than the CAD users. He comments on this result, observing that the knowledge required to operate a CAD system apparently interferes with the application of the knowledge required to make domain decisions. Contrary to what propaganda suggests, CAD does not unburden or "free" its user: Producing a drawing using CAD may be a more complicated and demanding task for a designer than drawing it manually.

Fricke (1999) observed that among his individually working designers, some who were successful "did not suppress first solution ideas" in their goal analysis, but "detached themselves during the conceptual design stage" from this early fixation (p. 428). They "produced variants," but reduced the number of suitable variants in critically evaluating them.

The apparent contradiction between these observations might be removed in at least two ways:

- Designers may declare or aspire to refrain from premature commitment, but in fact not put these intentions in practice (see also Malhotra et al., 1980; cf. our observation that designers' accounts about their activity often do not coincide with their actual activity, Visser, 1990).
- Early on in the design process, designers may select a kernel idea at a conceptual level, but later on, they may refrain from premature commitment at a more concrete or detailed level, by not fixing all values for its variables.

Time constraints and individual differences (cf. the apparently decisive role of Ball & Ormerod's, 2000, team manager) are two additional possible factors that may differentiate a designer's approach to the formulation of one or more solution options early, but also later on in the design process.

With respect to the first factor, Marples (1961), in his analysis of engineering-designers' decisions (a *post hoc* analysis, based on traces of their activity), concluded that under time pressure, the search for novel solutions was at discount, "to be avoided if possible." The solution proposals adopted as a starting point were thus "comparatively pedestrian and taken straight out of the stock-in-trade of the engineers concerned" (p. 65).

User Considerations
Guiding Development and Evaluation

As early as 1972, Rittel (1972/1984), in a discussion of planning societal problems (governmental planning, especially social or policy planning), states:

> There is no professional expertise that is concentrated in the expert [designer]'s mind. . . . The expertise used or needed, or the knowledge needed, in doing a design problem for others is distributed among many people, in particular those who are likely to become affected by the solution—by the plan—and therefore one should look for methods that help to activate their expertise. Because this expertise is frequently controversial, and because of what can be called "the symmetry of ignorance" . . . the process should be organized as an argument. (p. 320)

Rittel's expression "symmetry of ignorance" refers to his conviction that there is no reason, be it logical or educational, to defend that one "expert," that is, one particular participant in the whole decision process, knows better than another. Rittel considers this "symmetry of ignorance" a "nonsentimental" argument for user participation in design.

In user-centered system design, knowledge about the potential users of the system may guide both system development and system evaluation (Norman & Draper, 1986). Traditionally, system designers adopt task models as the reference for their specification of interactions between user and system (e.g., GOMS, see Card, Moran, & Newell, 1983).

Task analysis has undergone critical analysis, however. It has been argued, for example, that "(if used on its own) [it] would produce poor system designs because it fails to achieve sufficient device independence" (Benyon, 1992, p. 246). Task analysis has also been qualified as "pseudo-cognitive." It undeniably serves to identify and analyze users' tasks rather than their activities. Functional specifications for interactive systems may require, however, data on actual—and potential—activities.

More recently, scenario-based design (Carroll, 1995, 2000) and participatory design (or cooperative design) in general (Kyng & Greenbaum, 1991) are approaches used to gain data on user activities, aiming to take into account such data in to-be-designed systems.

"Participatory design" is "a term that refers to a large collection of attitudes and techniques predicated on the concept that the people who ultimately will use a designed artifact are entitled to have a voice in determining how the artifact is designed" (Carroll, 2006, p. 3).

Hollnagel (2002) defends a different position, which he qualifies as "function-centered" design. He considers that "human-centred design," which he sees as characterizing the second of three consecutive periods since the beginning of ergonomics, is "just as inadequate as machine-centred design," characteristic of the first period. The design of this period that adopted a "technological bias" toward the human-machine system, failed "because the growing complexity of systems made a decomposition approach to design inadequate." Hollnagel considers, however, that the period that followed has a "humanistic bias" and is just as inadequate. "It implies a dichotomy where one part of the system is seen as opposed to the other" and "suffers from being defined in a negative fashion in terms of what is not user-centred, rather than in terms of what is." Hollnagel proposes that system design adopt a "function-centred view where the focus is the joint cognitive system."

Software-system usability testing and more general evaluation methods of interactive systems can proceed by laboratory user testing or by more rapid techniques, referring to guidelines or other lists of inspection points (such as cognitive walkthroughs, heuristic evaluations, or even rapid ethnography; Millen, 2000).

As far as we know, little empirical research has been conducted on the way in which a designer's activity is guided by user considerations. Both Adelson et al. (1985) and Visser (1987b) observed designers to be guided, more or less, by user considerations. The strategy had, however, different functions in each study. In the first one, mental simulation of a user's interaction with the system helped designers to think of elements to be included in the design. As described in the following example, for the software engineer observed by us, ease of use for future users played a role in both development and evaluation of the software.

Example. The observed software engineer designing IPC software (in our software-design study, see Table 6.1) tried to take into account several types of users who were going to work with the IPC that he was designing (Visser, 1987b). He judged the system operators in the workshop as its most frequent users. Maintenance personnel were considered another category of IPC users. The engineer was observed to take into consideration these two types of users. Considering homogeneity of representation an important factor of ease of use, he used it as a design constraint, trying to make the software as homogeneous as possible, both for comprehension and for maintenance reasons. This search for homogeneity was realized in various ways, using the homogeneity constraint both in development and in evaluation of the software.

Using it as a generative constraint, the software engineer created uniform structures at several levels of the software: He homogenized the order of the lines of software in the modules, the branch order in the instructions, and the order of variables in the instruction branches.

Evaluating his software against the criterion of homogeneity sometimes led the engineer to reconsider instructions. In doing so, he modified either the current instruction in making it replicate the structure of previously written ones, or previously written instructions by giving them the structure of the current instruction.

One may suppose that the reason why such strategic consideration of users has not been observed in more studies is its possible dependency on real work situations. However, it was also noticed by Adelson et al. (1985) in a restricted laboratory setting.

21.6. DESIGN REPRESENTATION CONSTRUCTION ACTIVITIES

Many recent studies concern representational structures in design, especially external representations, but the cognitive activities and operations involved in their construction and use have not been the object of much research. Publications mention activities such as "transformation," "(re)interpretation," and "restructuring and combining." Generally, they describe the results that are obtained, but rarely make explicit the underlying cognitive activities or operations.

We distinguish three types of activities on representations, namely generation, transformation, and evaluation. These activities, their underlying operations, and related activities and operations are discussed in the following subsections. In special subsections, we review the use of knowledge in design, and specific aspects of collaborative design.

Problem Representation, Solution Generation, and Solution Evaluation: Three Stages in Design as Problem Solving

From a problem-solving perspective, design has often been described as proceeding through three stages, namely construction of problem representations, solution generation, and solution evaluation. A related, less high-level model sees these stages occurring in iterative cycles that, progressively, lead from the abstract, globally specified problem to its concrete, detailed implementable solution. None of these two models renders the actual design activity. The three stages correspond nevertheless to fundamental design activities, which are completely intertwined—and not at all consecutive, as stages are supposed to be. The perspective we have adopted, namely to consider design as the construction of representations rather than as problem solving, leads us to consider these three activities as construction of representations, even if they may involve different types of input and output representations.

Using Knowledge in Design

Knowledge is a central resource in the construction and use of representations. The importance of knowledge holds for most professional domains, but it is of course particularly critical in an activity that essentially consists in representational activities. Design requires general, abstract knowledge and weak, generally applicable methods, but designers also need domain-specific knowledge and the corresponding strong, knowledge-intensive methods. We suppose that satisficing, for example, requires more domain-specific knowledge than does optimizing. This also holds for the exercise of creativity, which is so important in design. In addition, knowledge is a key element in the exercise of analogical reasoning—which may, in turn, be related to design creativity (but see Visser, 1996).

It may seem surprising that there has been so little explicit discussion of knowledge in this book. Its role has, however, been referred to at many places, all through the different chapters of this book, even if more or less implicitly or *passim*.

Such references were not frequent in the parts dedicated to the SIP and SIT frameworks. In the presentation of the SIP approach, they were very general, because this problem-solving view insists mainly on generic knowledge and weak methods. In the SIP approach to problem solving, one searches for solutions in the "problem space," going "from one knowledge state to another, until the current knowledge state includes the problem solution" (Simon, 1978, p. 276). We quoted an interesting observation by Simon (1973/1984), who conjectured that the degree of a problem's structuredness may depend on the size of one's "knowledge base" (cf. also Newell's related view on this issue). Simon also noted the role of knowledge in scientific discovery, given that recognition depends on one's knowledge.

In our presentation of the SIT view, knowledge did not play an important role neither, but for different, nearly opposite, reasons. SIT-inspired researchers have identified and described in detail much domain-specific knowledge. They insist on the role of "knowledge-in-action"—which they oppose to school knowledge, whose role is of course not denied, but ignored in their research. SIT-inspired studies have provided us with extremely rich descriptions of situations that were often so unique that presentation of the knowledge identified would have been rather anecdotal. One may notice that it is undeniably difficult to find a level of description of interest to many different people (researchers, practitioners, students, general public), with different backgrounds and interest in different domains. Furthermore, SIT-inspired researchers emphasize that there is more to design—and other professional practice—than knowledge (cf. Bucciarelli, 1988).

It is in this part, in which we present our own view that focuses on representations and their construction, that knowledge has often been present, but implicitly.

Yet, without knowledge, no representation! Knowledge is necessary—but of course not sufficient—for the construction of representations. Without knowledge, no interpretation, thus neither the possibility to look at a project in a way different from one's colleagues, nor that of seeing things differently than one did during a previous project! The operative and goal-oriented character of representation results from an interaction between one's knowledge and experience, and the situation one is in.

Nonalgorithmic activities—necessary in, for example, creativity, satisficing, (re)interpretation, and qualitative simulation—require knowledge. In order to proceed to complex calculations, a designer, of course, also needs knowledge, but of a sort that can be learned in school. The knowledge that is very important in design is not gained through formal education, but through experience. Designers may acquire such knowledge because of their work on many different types of projects, and their interaction with colleagues who have other specialties (see Falzon & Visser, 1989).

Knowledge determines if a design task constitutes a problem for someone. Working with ill-defined problem data is only possible if one has specific knowledge (in addition to generic knowledge, of course).

Furthermore, knowledge is a critical resource underlying most strategies. If simulation via representations works, it is thanks to one's knowledge. Reuse is, by definition, impossible without knowledge (it is not a components library that makes knowledge superfluous). Handling constraints (especially constructed constraints) would be hard without it.

The domains from which this knowledge comes are not only the application domain and that of design methods, but also the underlying technical and theoretical domains (mathematics, science, engineering)—and even nontechnical domains. In our carrying/fastening device study (Visser, 1995b), we showed the importance of commonsense knowledge (in the design project examined, this was the knowledge of cycling). Additionally, designers, one may hope, also draw on ergonomics and knowledge of social, political, economic, and legal aspects of the artifact and its use. As designers generally are not expert in all these different domains, the need of design projects for wide-ranging knowledge requires collaboration between professionals from various domains.

With respect to knowledge of different abstraction levels, designers of course use much generic, abstract knowledge (first principles, general-purpose knowledge, weak methods). However, the reuse of specific knowledge related to particular past design projects plays an essential role in design (Visser, 1995b). In our carrying/fastening device study (Visser, 1995b), we observed how the knowledge of cycling is not theoretical, school knowledge, but the result of personal experience in cycling, with or without a backpack, on a mountain bike or other bicycle. We showed how this episodic knowledge (Tulving, 1972, 1983) grounded in personal experience may be used in various ways (both in the construction of representations used for the generation of solution ideas, and in the evaluation of solution proposals). In this study, we also showed the importance of human informants besides non-human information sources. We observed how designers often use colleagues as informants—and how colleagues present themselves as such without being requested explicitly (Berlin, 1993; Visser, 1993a).

These are only a few examples, mentioned in order to indicate the importance of knowledge in design.

Expertise and knowledge. We did not discuss specifically the role of expertise—except in the context of the route-planning studies. Such a discussion would have involved many references to the differential possession and use of knowledge—though knowledge is far from being the only element that distinguishes experts from novices. We saw indeed in the route-planning research that the possession of knowledge plays a decisive role in the choice of strategies adopted and the selection, adjustment, and integration of constraints.

There are at least three types of research on expertise. The comparison between experts and novices in a domain, that is, studies on *levels* of expertise, is the classical paradigm in studies on interindividual differences in this domain (Chi, Glaser, & Farr, 1988; Cross, 2004a, 2004c; "Expertise in Design," 2004; Glaser, 1986; Glaser & Chi, 1988; Reimann & Chi, 1989). Experts have also been studied in clinical studies, leading researchers to identify particular characteristics of particular experts (Cross, 2001b, 2002).

We have proposed to distinguish also different *types* of expertise (Falzon & Visser, 1989; see also Visser & Morais, 1991). We analyzed how experts in the same domain may exhibit different types of knowledge, and observed that this knowledge is also organized differently between the experts. We attributed these differences to different task experience (workshop vs. laboratory in the context of the aerospace industry). Our analysis of previous studies by colleagues who compared experts showed, in addition to the role of one's task, the importance of the representation that one constructs of one's task. The comparison between the two experts examined led us to qualify the knowledge of one expert's as "operative" and that of the other as "general." "The two experts differ in the same way as a teacher differs from a practitioner, in the same way as an epistemic subject differs from an operative subject" (Falzon & Visser, 1989, p. 125).

Generation and Transformation of Representations

In the coding scheme for his sketching study, Goel (1995, p. 207) considers that the source of a diagram can be a previous solution, or one's memory (i.e., here, long-term memory, LTM). He codes the source as memory if, syntactically, the diagram is "a new drawing" and semantically, it translates "a new idea."

Goel (1995) distinguishes two "transformation types of diagram generation," namely "new generation" and "transformation." "All (and only) syntactic and semantic source types that have their origin in LTM are . . . new generations" (p. 210). New generation receives little attention relative to lateral and vertical transformation. "Transformation" (which we qualify as "transformation*" in Table 21.4) can take three forms: Duplication, lateral (or "divergent," Rodgers et al., 2000) transformation, and vertical (or "convergent," Rodgers et al., 2000) transformation (see Table 21.4).

We wish to emphasize that "transformation" will seldom be the transformation into a well-defined terminal state or object, through a well-defined process, program, or sequence of operations—such as is the case for the classical "transformation problems" in cognitive psychology. With respect to our qualifying "generation" and "transformation" as the main design representation-construction activities, one may remember Reitman (1964), who takes together, in his analysis of different types of problems, all those that involve the "transformation or creation of states, objects or collections of objects" (p. 284).

Generation. A representation is never generated "out of nothing" (*ex nihilo,* from scratch). We consider it difficult, if not impossible, to decide if an idea or drawing (or other representation) is "new." In accordance with Goel, who considers both "new generation" and "transformation" as types of *transformation,* we consider that design always consists in *transformation* of representations. We qualify the construction of representations as "generation" (Goel's "new generation") if its *main* source is one's memory—something that will be difficult to observe for an external observer. We insist on "main," because memory will never be the only source. By definition, the state of a design project (requirements and their follow-up included) will influence a designer. In addition to this influence, there will be other contributions "from the outside world."

Designers will interpret the input to a design project, that is, the requirements and other data that they receive or collect (e.g., reference documents, similar artifacts), in order to generate a first representation—which may consist of an ensemble of representations: For example, one or more related mental and external representations.

Generation may be implemented by different types of processes and operations: From the "simple" evocation of knowledge from memory to the elaboration of "new" representations out of mnesic knowledge entities without a clear link to the current task (e.g., through analogical reasoning and other nondeterministic leaps; Visser, 1991b).

TABLE 21.4
Our Version of the Two Types of Transformation
and Three Forms of Transformation Proposed by Goel (1995)

Transformation types of generation		
new generation		
transformation*	*Forms*	lateral transformation
	of	vertical transformation
	*transformation**	duplication

Note. For reasons of intelligibility, we use "transformation*" to refer to one of the two "transformation types" proposed by Goel (1995).

This evocation-versus-elaboration opposition is comparable to the one that Newell (1990) establishes between "generation by knowledge retrieval" (or "preparation") and "generation by search" (or "deliberation").

The distinction between generation by evocation and by elaboration of course does not correspond to a clear-cut opposition, but is an analytical distinction that refers to a continuous dimension. Elaboration of a representation always uses mnesic entities, which will have been evoked from memory (via "unguided" memory activation or "guided" memory retrieval mechanisms, Visser, 1991b). We have illustrated this idea elsewhere by observations from our composite-structure design study (Visser, 1991b).

Schema instantiation is a form of knowledge evocation that has received much attention in software-design studies. Schemata have indeed been the main framework for the analysis of knowledge representation in cognitive software-design research (Détienne, 2002a).

Generation of representations may use operations and other activities, such as information gathering.

Transformation. We propose to distinguish transformation activities according to the type of transformation between input representation r_x and output representation r_y. We distinguish the following forms. Transformation activities may

- *duplicate* (Goel, 1995), that is, replicate or reformulate r_i.
- *add*, that is, introduce new information or "small alterations" (Van der Lugt, 2002) into r_i.
- *detail*, that is, break up r_i into components r_{i1} to r_{in}.
- *concretize*, that is, transform r_i into r_i' which represents r_i from a more concrete perspective.
- *modify*, that is, transform r_i into another version r_i', neither detailing, nor concretizing it.
- *revolutionize* (or *substitute*, Visser, 2006), that is, replace r_i by an alternative representation r_j, neither detailing, nor concretizing it (corresponding to Van der Lugt's, 2002, "tangential transformations," i.e. "wild leaps into a different direction").

Even if Goel (1995, p. 119) states that lateral transformations lead to "slightly different" versions, we consider that both transformations into different versions (through modifying) and into alternative representations (through revolutionizing) constitute "lateral" transformations. Goel's (1995, p. 119) vertical transformations (which, in his terms, introduce "details") cover what we call both "detailing" and "concretizing."

Many activities play a more or less direct role in these different types of transformation. Some examples (varying between operations and activities) are interpretation, association, brainstorming, reinterpretation, confrontation, articulation, integration, analysis, exploration, inference, restructuring, combining, drawing (sketching, drafting, and other forms), hypothesizing, and justifying. In this book, we comment on only some of them.

We do not discuss here activities that correspond to configurations of other activities, such as cognitive synchronization, or conflict resolution, which we analyzed in terms of "exchanges" (polylogal[33] verbal-interactional units) in our analysis of technical-review meetings (D'Astous et al., 2004). Neither do we present basic processes (mechanisms). Finally, we do not consider activities at the action-management level, such as planning, organization, regulation, or control.

Even if it is too simplistic to qualify "analysis" as a first design "stage," analyzing indeed corresponds to a central activity in the initial phases of a design project. Constraints analysis is essential to disambiguate design requirements. Analyzing the current design state may be a way to introduce detail or concreteness in the project. "Analysis" has, however, a logical undertone, which causes that it can certainly not be the only—or even the main—activity in the initial design phases. Other, more nonalgorithmic activities will also be required, such as interpretation, association, brainstorming, and exploration.

Analogical reasoning occurs in all three representational activities. We have mentioned it in different contexts: As a factor of opportunism, in creativity-requiring activities, as a way to tackle ill-defined problems by interpreting them, and as a possible form to generate "interesting" design ideas. It is also the reasoning form that underlies reuse, which plays an important role in design.

We observed its role in different studies, several examples of which have been presented in this book. As an example of one of the six factors of opportunism that we identified, we described analogical reasoning used by the mechanical-design engineer in our functional-specification study (see Table 6.1). Using analogies, he took advantage of representations that he was constructing and using for his current design actions, to design analogically related design objects. A completely different use of analogy has been observed in the composite-structure aerospace design study (see Table 6.1). There, the designer especially employed analogical reasoning in the conceptual-design stage. When elaborating conceptual solutions to design problems, he and his colleagues frequently were observed to be reminded of extradesign domain objects that implemented concepts (principles, mechanisms) that they judged potentially useful for development of a solution to the current design problem. The following example (from Visser, 1996) illustrates this use of analogy that we analyzed as contributing to the innovative character of the design project (other examples are presented in Visser, 1991b).

Example. When the composite-structure designer and his colleagues are developing, in a discussion, "unfurling principles" for antennas, they come up with ideas such as an "umbrella" and other "folding" objects. They proposed, for

---

[33] A "polylogue" is a dialogue between more than two participants.

example, a "folding photo screen," a "folding butterfly net," and a "folding sun hat," all related to the target by analogical relationships.

Different forms of inference are of course also used in design. Induction is used much more frequently than deduction. Goel (1995) identifies only 1.3% "(overt) deductive inferences" in his observations. In our composite-structure design study (see Table 6.1), neither did we notice any overt form of deduction.

The articulation, combination, and integration of representations play a particular role in collaborative design. So do inform, comment, and request. Such activities are discussed in the subsection *Construction of Interdesigner Compatible Representations.*

Restructuring and combining representations are often mentioned as components of the creative process (Verstijnen et al., 1998; Verstijnen et al., 2001). Verstijnen et al. show that restructuring and combining are two separate constituents of creativity that function differently. In distinct ways, each can lead designers to introduce new information in the current design representation — something that is useful in generation and transformation of representations.

Restructuring is qualified by the authors as "getting free from an original conception" (Verstijnen et al., 1998, p. 545). Verstijnen et al. claim that "mental imagery" operations (i.e., operations on mental images) may lead to discovery of new ideas — but only under certain conditions. Some operations cannot be performed "within mental imagery alone" and other operations "are much easier to perform externally" (p. 522).

It is difficult to restructure completely mentally an existing external representation (i.e., a drawing, in Verstijnen et al.'s experimental studies) — for novices, it is even impossible. It is facilitated if one is allowed or encouraged to sketch — but this facilitation only holds for experienced designers. However, combining (synthesizing) parts of a representation can be performed mentally by only using mental imagery. In that case, "no additional value is obtained from sketching" (Verstijnen et al., 1998, p. 535; cf. the hypothesis formulated by Kovordányi, 1999, that mental images are difficult to reinterpret because "visual selective attention may inhibit the emergence of new interpretations"). One may indeed suppose that the two operations — restructuring and combining — impose different loads on mental processing.

Yet, inventors (such as Kekulé, an example presented by Verstijnen et al.) seem to be able to restructure exclusively "in their head." Verstijnen et al. (1998) formulate the interesting hypothesis that "extraordinarily creative individuals" may be able to "construct analogies within imagery, for which others, in more mundane cases, require a sketch" (p. 546). Indeed, what an external representation such as a sketch allows a person is to restructure their image (i.e., an internal representation) in analogy to that external representation. This inspires in Verstijnen et al. (2001) the idea that "with no paper available or no expertise to use it, analogies can be used to support the creative process instead of sketches" (p. 1) — but perhaps only in "extraordinarily creative individuals" (the addition is ours).

As tools for reinterpretation, activities such as restructuring and combining may thus be used to come up with new ideas. Drawing (i.e., sketching, drafting, and other forms of drawing) may also be a tool for other activities. Besides restructuring, it may serve, for example, analysis, and simulation. It may also fulfill interactional functions, such as informing or explaining. It can even have several functions simultaneously: For example, simulation, explanation, and storing. The relatively unstructured, fluid, and imprecise drawings that sketches are, may give access to knowledge not yet retrieved and may evoke new ways of seeing (because of their non-notionality; see Goel, 1995). Unforeseen views on the design project in progress are supposed to open up unanticipated potentialities for new aspects or even completely new directions.

Evaluation of Representations

According to design methodologies, the generation and evaluation of solutions are two different stages in a design project. Many empirical studies have shown, however, that designers intertwine the two. The participants in the technical review meetings that we studied (D'Astous et al., 2004) were supposed to follow a particular method in which design was not supposed to occur. They came up, however, with alternative solutions; that is, not only were they recording the underlying negative evaluations, but they also proceeded to design.

Evaluating an entity consists in assessing it vis-à-vis one or more references (Bonnardel, 1991a). In the context of design, evaluation may occur when a representation is presented by its author, or interpreted by colleagues, as an "idea" or "solution proposal." Colleagues may interpret a representation as a solution proposal without its author presenting it explicitly as such, and they may evaluate it without its author explicitly requesting them to do so (Visser, 1993a).

Terminology. The terminology around "constraints" and "criteria" is still under debate in the domain of cognitive design studies. Bonnardel (1989) reserves the term "constraints" for operative evaluative references and "criteria" for conceptual references, whereas we use "constraints" for generative references that steer solution generation and "criteria" for critical evaluative references guiding solution evaluation (Visser, 1996). Other distinctions have also been proposed.

According to the source of an evaluative reference, researchers distinguish different types of evaluative references (Bonnardel, 1991a; Ullman et al., 1988):

- Prescribed constraints, which are given to the designer or which the designer infers from the problem specifications.
- Constructed constraints, for which designers mainly use their domain knowledge.
- Deduced constraints, which designers infer based on other constraints, the current state of the design project (the problem solution), and design decisions made during their design problem solving.

Depending on the type of reference used by a designer, researchers distinguish three evaluation strategies (Bonnardel, 1991b; Martin, Détienne, & Lavigne, 2000; Martin et al., 2001), all three of which we qualify as "comparative":[34]

- Analytical evaluation: A solution is assessed vis-à-vis a number of constraints.
- Comparative* evaluation: Various solution versions or alternatives are compared with each other.
- Analogical evaluation: A solution is assessed using knowledge acquired in relation to a previous solution.

In an analysis of negotiation patterns between participants in multidisciplinary aeronautical-design meetings, Martin et al. (2000, 2001) show that if such evaluation does not lead to a consensus between the different partners, arguments of authority may be used.

Evaluative references are forms of knowledge. As expected, designers' expertise in a domain influences their use of these references (D'Astous et al., 2004).

Given that, in a collaborative design setting, designers may have different representations of a project, proposals are evaluated not only based on purely technical, "objective" evaluative criteria. They are also the object of negotiation, and the final agreement concerning a solution also results from compromises between designers (Martin et al., 2000, 2001). In addition, not only solution proposals, but also evaluation criteria and procedures undergo evaluation (D'Astous et al., 2004).

The preceding discussion concerned different forms of evaluation by comparison, that is, with respect to evaluative references. This type of evaluation is possible if the form of the representation that is to be evaluated allows such a comparison. For example, if one knows already the performance measures of the artifact. This is often the case in engineering, where "objective" measures of artifacts are possible (e.g., measures of their future performance).

The evaluation of other types of artifacts may be based on simulation. The result of such simulation (e.g., a certain behavior displayed by the artifact) may constitute the input of comparative evaluation.

Evaluation has functions at both the action-execution and the action-management level of the design activity. The classical solution evaluation occurs at the action-execution level and leads generally to the selection of one proposal—possibly after one or more iterations. At the action-management level, evaluation affects the progress of the design process. Depending on the results, design may be pursued in different ways. Designers, thus, evaluate not only solutions, but also their possible design process, its progression, and direction (Visser, 1996).

---

[34] We therefore label the specific evaluation mode that is called "comparative evaluation" in the literature, as "comparative* evaluation" in order to distinguish it from the more general "evaluation by comparison," of which it is only one particular form and that also includes the analytical and analogical evaluation modes.

Collaborative Design Through Interaction

Collaborative design takes different forms and refers to the various representation-construction activities presented earlier. Besides the functions that representations play in both individual *and* collective design settings (mainly cognitive offloading, reminding, keeping track, storage, communication, organizing, reasoning, and discovery), various aspects of the externalization possibilities of representations provide additional functions specific to collective design. These functions go together with different cooperative activities, which vary according to the phases of the design project. During distributed design, when the designers' central activity is coordination in order to manage task interdependencies, representations of course play a role. Yet, it is in co-design that they have a particular function, due to its collaborative setting.

In collaborative-design situations, individual design plays of course also an important role (as we have emphasized at different occasions in this book; see also Visser, 1993a, 2002a). Yet, an essential part of collaborative design, especially during co-design, takes place—that is, advances—through interaction. This apparently unequivocal statement—it may even seem tautological—conveys characteristics of design thinking that we consider essential.

Indeed, the different forms that interaction may take in collaborative design—especially, linguistic, graphical, gestural, and postural—are, in our view, not the simple *expression* and *transmission* (communication) of ideas previously developed in an internal medium (such as Fodor's "language of thought"). They are more and of a different nature than the trace of a so-called "genuine" design activity, which would be individual and occur internally, and which verbal and other forms of expression would allow sharing with colleagues. On this issue, we do not concur with Goldschmidt (1995) when she writes that "thinking aloud and conversing with others can be seen as similar reflections of cognitive processes," which we can accept as "equal windows into the cognitive processes involved in design thinking" (p.193).

Notice that, in these collaborative contexts, a fundamental part is played by other factors than cognitive ones (representations, knowledge). These are especially emotional factors, and social, institutional, and interactional factors, such as the roles of the different design participants (formal, static roles that depend on one's predefined function in the design project, and informal roles that emerge and evolve depending on the interaction; see D'Astous et al., 2001; Fagan, 1976; Herbsleb et al., 1995; McGrath, 1984; Seaman & Basili, 1998).

In the following and last two divisions of this section, we discuss activities that are specific to collaborative design.

Construction and Use of Intermediary Representations in Collaborative Design

Many notions referring to the interdesigner intermediary function of representations in collaborative design have been proposed in the literature, such as "intermediary objects," "coordinative artifacts" (Schmidt & Wagner, 2002), "entities for cooperation" (Boujut & Laureillard, 2002), and "boundary objects" (Star, 1988, discussed later).

Emphasizing the material setting and the artifactual nature of these entities that are essential in designers' interaction, Schmidt and Wagner (2002) emphasize that, in cooperative work, their main role is not informative, but coordinative: They contribute to a more or less effortless and fluent coordination and integration of individual activities in coordinative practices.

For architects, a particular form of coordinative artifacts is "layered artifacts." They are a tool that architects use "for communicating things that need to be taken account of or changed." Schmidt & Wagner (2002) describe that architects construct them by

> making annotations on a document, e.g., putting a red circle around a problem, adding details (correct measures, material), marking a part of a drawing with a post-it with some instructions for changes, corrections (e.g., in pencil directly on a plan), sketching either directly on a plan copy or on transparent tracing paper. . . . Layered artifacts facilitate coordination between activities (and the people who are responsible for them). They, for example, provide a collective or individual space for experimentation and change. The CAD drawing itself is a layered artifact, which builds on a particular mix of codes for functions and materials and has been tailored to a particular division of labor. (pp. 10–11)

The benefit of visual expression in creative collective activity has been examined by Van der Lugt (2002). One of the supposed specific contributions of visual expression to idea generation in a collective setting is that, through conversation with the drawings of colleagues, people may build on each other's ideas. Van der Lugt shows that sketching using brainsketching tools indeed contributes to creative activity in idea-generation groups, but not as expected: It especially supports reinterpretation of one's own ideas, and so stimulates creativity in *individual* idea generation. Reinterpretation of ideas generated by other group members is not enhanced. Collective working is thus not the panacea for all complex processes. Individually conducted activities in collective settings may sometimes lead to "better" results. The visual expression in a collective setting may nevertheless improve integration of the group process, by facilitating the access to previously expressed ideas.

Van der Lugt emphasizes that his results may be specific to the techniques and tools examined, and thus cannot be generalized to other sketching and idea-generation tools. Indeed, in another study on sketching tools, using a different technique (visual brainstorming), Van der Lugt (2000) observed a breakdown in the idea-generation process.

Another communicative situation in design projects is the interaction between people involved in design and in implementation. Eckert's study of knitwear design (presented in Stacey & Eckert, 2003) constitutes an interesting example of the difficulties that these situations may bring about.

The knitwear designers examined by Eckert use "technical sketches" in order to communicate their patterns and garment shapes to the machine technicians who are to implement the knitwear designs in garments. In addition to a free-hand drawing part (the actual sketches), these documents comprise a short verbal description and a set of dimensions (Stacey & Eckert, 2003, p. 157). These technical sketches are supposed to clarify the designer's specifications, but "are often excessively imprecise or ambiguous." The technicians tend to ignore the actual sketch part, and "rely mainly on the verbal descriptions, which only give broad indications of categories" (pp. 157–158). The technicians are not able to distinguish in these documents the important and relatively exactly specified design aspects from unimportant details and elements that are placeholders for broad categories (e.g., the type of neckline or the chest pattern). As they "have no way of judging what to believe, [they] usually take what is standard as more likely to be reliable" (p. 174). This leads to the products, that is, the garments, often being more traditional than intended by their designers. The technicians repeatedly produce garments "that violate the designers' intentions." They also often state that "what the designers want can't be done" (p. 174).

Notice that this conclusion—technicians refer to standards for understanding the specifications they receive—is not restricted to these specific technicians and these particular technical sketches. It may hold for anybody who is to interpret any semiotic expression produced by other people. Both designers and other participants in the development process of an artifact, interpret the language as well as the graphical expressions by their colleagues, in terms of the standards they are familiar with—and of their own past experience of artifacts more or less similar to the current object of the design project.

Still another communicative situation—but one that is not necessarily present in every design project—is that between designers and users. With respect to interactive-software design, Carroll (2006) notices that there is a big and crucial "gap" between the worldviews held by designers of software and its potential users. Participatory design is one way to bridge this gap. Research in this domain has produced many proposals for possible design representations enabling the two parties to communicate:

> Many of these approaches essentially implement a user interface design at the earliest stage of system development: Designers can show concretely what they have in mind, rather than specifying it mathematically, and other stakeholders can react and critique what they can actually see and manipulate. . . . A slightly more abstract approach is scenario-based design in which system functionality and the experience of using that functionality are described in narrative episodes of user interaction. (p. 11)

Argumentation—a "hot item" in studies on cooperative activities—has only been touched on in this book (cf. Rittel's, 1972/1984, argumentative model). Authors attribute a more or less broad sense to the notion. We conceive argumentation as an attempt to modify the representations held by one's interlocutors. Many activities in co-design are thus argumentative.

Boundary representations. As advanced by Star (1988), in collaborative design, one needs "boundary objects" to serve as an interface between people from different "communities of practice." These objects may take many artifactual forms, for example, representational. We have proposed to qualify as "boundary representations" (no connection to the b-rep model for representing a cube) the representational version of boundary objects (Visser, 2006). The fact that they work does not mean that partners from different communities view or use them in the same way. Different partners may interpret them differently, but they work if they contain sufficient details understandable by these parties. No party needs to understand the full context of use adopted by their interaction partners. It is the acknowledgment and discussion of the differences that enable people to use them successfully together.

An example of a representation meant as a boundary representation is the technical sketch used by the knitwear designers examined by Eckert (see Stacey & Eckert, 2003, p. 163, see also our earlier presentation). They do not work as boundary objects, because they do not contain sufficient detail to be understandable by the different parties involved. Successful communication depends not only on "the sender's use of appropriate representations for information," but also on "the recipients' ability to construct meaning from those representations" (Stacey & Eckert, 2003, p. 158).

According to Stacey and Eckert (2003), two factors play herein a particularly important role:

> The extent to which the participants share context and share expertise; and the tightness of the feedback loops. . . . In face-to-face communication, failures of comprehension can be identified and conveyed very quickly, and speech, gestures and sketches are used to explain and disambiguate each other. . . . In less tightly coupled exchanges, the need to prevent rather than correct misunderstanding is correspondingly greater. (p. 162)

With respect to these factors, Eckert observed that in nearly all companies that she had visited, designers do their conceptual design without any input from the technicians that are to implement their designs. This absence of communication may explain, at least in part, that the technical sketches used as specification documents by the knitwear designers are ambiguous—that is, in the form that the two parties use them: Without any other interaction allowing them to be acknowledged and discussed. "The less the participants discuss, and the less knowledge and contextual information they share, the more sketches, diagrams and other communications need to carry with them the means of their own interpretation" (Stacey & Eckert, 2003, p. 163).

Construction
of Interdesigner Compatible Representations

In a paper on "bringing different points of view together," Fischer (2000) writes:

> Because complex problems require more knowledge than any single person possesses, communication and collaboration among all the involved stakeholders are necessary; for example, domain experts understand the practice, and system designers know the technology. Communication breakdowns are often experienced because stakeholders belonging to different cultures (Snow, 1993) use different norms, symbols, and representations. Rather than viewing this *symmetry of ignorance* (Rittel, 1984) (or "asymmetry of knowledge") as an obstacle during design, we view it as an opportunity for creativity. The different perspectives help in discovering alternatives and can help uncover tacit aspects of problems. (p. 3)

Construction of interdesigner compatible representations when co-designing proceeds through activities qualified as "grounding" (Clark & Brennan, 1991) and "cognitive synchronization" (D'Astous et al., 2004; Falzon, 1994), through a negotiation process resulting in "social constructions" (Bucciarelli, 1988) or through argumentation resulting in the settling, "dodging," or substitution of "issues" (Kunz & Rittel, 1979). A great amount of time is spent on these activities (D'Astous et al., 2004; Herbsleb et al., 1995; Karsenty, 1991a; G. M. Olson et al., 1992, 1996; Stempfle & Badke-Schaub, 2002). Recent studies have observed that synchronization can also take a gestural form (cf. research in Tversky's STAR team, http://www-psych.stanford.edu/~bt/gesture/, retrieved August 16, 2005).

In our study on software-review meetings (D'Astous et al., 2004), we showed that the construction of interdesigner compatible representations of the to-be-reviewed design solution was a prerequisite for the occurrence of evaluation activities, which were the prescribed task. We also observed that cognitive synchronization concerned not only the problem solutions but also the criteria and the evaluation procedures.

Given that designers have their personal representations, collaboration between designers calls for confrontation, articulation, and integration of these different representations, in order for the designers to be able to reach a solution that is adopted for the common activity. The confrontation of personal representations also leads to conflicts between designers, which they are to resolve (see a remarkable early study in the domain of architectural design by M. Klein & Lu, 1989).

An interesting reading of Simon's (1969/1999b) thinking about representations is provided by Carroll (2006). Carroll notices that in the second edition of *Sciences of the Artificial*, Simon's view seems changed. In "Social Planning," a new chapter in this edition, Simon "suggested that *organizations* could be considered design representations (pp. 141–143), using the example of the Economic Cooperation Administration (ECA), the entity that implemented the Marshall Plan in 1948" (p. 12). At the outset, people involved in ECA did not agree on this agency. Carroll quotes Simon who "observes (p. 143), 'What was needed was not so much a 'correct' conceptualization as one that would facilitate action rather than paralyze it. The organization of ECA, as it evolved, provided a common problem representation within which all could work'" (p. 12). As the ECA proceeded, one of the six original conceptions prevailed. Carroll comments, "many uses of prototypes in participatory design are compatible with this suggestion; prototypes provide an evolving framework for exploring design options and gradually focusing on a final solution" (Carroll, 2006, p. 12).

22 Discussion

Our approach to design is discussed in the general *Conclusion*. Here we only discuss two specific points to which we do not return. We first consider the assumption that certain dimensions of the object of design influence the design activity: In spite of the validity of the "generic design" view—that is, design is a specific cognitive activity, distinct from other types of cognitive activity —, there are different "forms" of design. The second discussion issue prolongs the object of this chapter. After identification and characterization of the representational structures in design and the activities operating on them, the next—and necessary (Anderson, 1978)—step is to link the two in an integrated cognitive model of design. Based on tendencies identified in cognitive design studies, we outline here some initial directions regarding functional linkages between representational structures and activities.

22.1. DESIGN AS A SPECIFIC COGNITIVE ACTIVITY, IN SPITE OF DESIGN TAKING ALSO DIFFERENT "FORMS"

"Activities as diverse as software design, architectural design, naming and letter-writing appear to have much in common" (Thomas & Carroll, 1979/1984, p. 234).

Even if often not stated explicitly, the tendency in much cognitive design research is to consider design a specific cognitive activity. As shown by Goel and Pirolli (Goel, 1994; Goel & Pirolli, 1989)—but in restricted experimental conditions, not in real, professional design projects—, besides important similarities between design activities in different domains, there are also important differences between design tasks and nondesign tasks (see also Thomas & Carroll, 1979/1984). Based on this double observation, a "generic design" hypothesis has been advanced: Designing is a cognitive activity whose implementation (a) in design tasks in different application domains has strong commonalities and (b) has distinctive characteristics from other cognitive activities. The hypothesis has not been substantiated, however, through comparative cognitive analyses, except by Goel and Pirolli (Goel, 1994; Goel & Pirolli, 1989) in their experimental work.

We adhere to this hypothesis. Yet, we suppose that the nature of the artifact to be designed and the situation in which it is being designed introduce specificities in the corresponding activities and consequently lead to different "forms" of design. We therefore advance the hypothesis that, in spite of the validity of the generic-design idea, designing may take different "forms" depending on these factors. Our hypothesis takes the following form:

(1) Design thinking has distinctive characteristics from other cognitive activities.

(2) There are strong commonalities between the implementations of design thinking in design tasks in different application domains.

(3) There are also differences between these implementations of design thinking in different domains.

(4) Nevertheless, these differences do not reinstate strong commonalities between designing and other cognitive activities, whereas the commonalities between all the different forms of design thinking are sufficiently distinctive from the characteristics of other cognitive activities, to consider design a specific cognitive activity.

In this section, we are only concerned with the third point and introduce material to support it. The first and the second points have been discussed in detail by Goel and Pirolli (Goel, 1994; Goel & Pirolli, 1989). The fourth point, however, remains completely hypothetical and calls for new empirical research comparable to that conducted by Goel and Pirolli, but in real design situations.

Our discussion begins, once again, by a reference to the position that Simon—implicitly—adopted with respect to design in different domains. If Simon adhered to the hypothesis that we are formulating here—at least, to its third point—this could explain, at least in part, his different views on standard design (exemplified by engineering and architectural design) and social planning (and possibly, inventive engineering design). His publications, however, did not touch on this idea.

It is mostly in informal discussions, and without presentation of empirical or theoretical evidence, that this issue has been hinted at (Löwgren, 1995; Ullman et al., 1988). The engineering-design methodologists Hubka and Eder (1987), for example, assert that "the *object* of a design activity, what is being designed, . . . substantially influences the design process" (p. 124).

In this section, we propose some candidates for dimensions underlying differences between forms of design. Our discussion is often only allusive. This section presents material that needs to be further analyzed and developed. We present here a number of possible directions, for other researchers to follow—or modify—and complete them.

As noticed explicitly in certain later subsections, interindividual differences may often affect the influence of a dimension. Certain designers may be more inclined to use certain types of representations or other tools, or be more at ease with their use.

Maturity of a Domain and Available Design Tools

The NSF "Science of Design" program (*Science of Design*, 2004) aims to "develop a set of scientific [design] principles to guide the design of software-intensive systems." An underlying idea is that

in fields more mature than computer science [such as architecture and other engineering disciplines, such as civil or chemical engineering], design methodology has traditionally relied heavily on constructs such as languages and notational conventions, modularity principles, composition rules, methodical decision procedures and handbooks of codified experience.... However, the design of software-intensive systems is more often done using rough guidelines, intuition and experiential knowledge.

Even if cognitive design research has shown the difficulty of designers' effectively working according to design methodology prescriptions, one may suppose that being familiar with the constructs and other tools that have been developed in a domain, may influence—one supposes, facilitate—designers' activity.

Especially in the domain of software design, research has shown that design methodologies may have an influence on the design activity and on the resulting artifact (Lee & Pennington, 1994, for object-oriented and procedural paradigms).

Use of Internal Versus External Representations

As mentioned earlier, according to Zhang and Norman (1994), internal and external representations activate different types of processes, that is, cognitive or perceptive. We also noticed that, in our opinion, things are not so systematic. Nevertheless, one may expect that the use of internal and that of external representations involve processing differences. Therefore, designing may differ between domains depending on the importance of external representations in these domains.

With respect to this dimension, we wish to state explicitly that undeniably its importance depends also on the designer.

Use of Different Types of External Representations

This point particularizes that of the previous subsection.

One of the factors underlying the differences that seem to exist between software and other types of design (see later discussion) may be due to the different types of external representations primarily used. The possibilities provided by particular types of external representations compared to those offered by alphanumerical representations (especially, with respect to the ease of visualization and manipulation, and their corollaries) may facilitate, for example, simulation and other forms of evaluation.

This observation probably applies not only to classical (i.e., nonvisual) forms of software design. It may also hold for the design of other symbolic artifacts, such as other procedures, plans, and organizational structures.

Dependencies Between Function and Form

This distinction opposes domains where function and form can be aligned, to domains where individual forms may perform many functions. In the first type of domain, to each particular form corresponds a particular function. In software design, for example, to a functional decomposition may correspond more or less directly a structural decomposition. In the second type of domain, for example, mechanical design, each design decision can affect each subsequent decision, because a goal may be achieved, structurally, by modifying a previously specified form rather than by introducing a new one (Ullman et al., 1988). One may suppose that these differences bring about differences in designers' decomposition activities.

"Designing in Space Versus Time" (Thomas & Carroll, 1979/1984)

As discussed previously (see the subsection *Temporal and Spatial Constraints*), several studies have shown that designers deal differently with temporal and spatial constraints (Chalmé et al., 2004; detailed in Visser, 2004; see also Thomas & Carroll, 1979/1984). Research has not yet settled clearly the relative difficulties involved in the different types of design that preferentially implement these two types of constraints. It has even less identified the underlying factors.

Design of Structures Versus Design of Processes

This dimension is related to the previous one. Structures (which may correspond to states) are not necessarily spatially constrained, but processes have systematically temporal characteristics. By analogy to the differences between the cognitive treatment of spatial and temporal constraints, one may expect that structures and processes are represented differently (especially mentally,[35] but also externally), thus processed differently, and therefore lead to different design activities (cf. Clancey's, 1985, distinction between configuration and planning).

Distance Between Design Concept and Final Product

Löwgren (1995, p. 94) opposes "external" software design ("design of the external behavior and appearance of the product, the services it offers to users and its place in the organization") to other types of design, for example, architectural and engineering design. In external software design, "it is technically possible to evolve a software prototype into a final product"—something difficult (or even impossible) in these other fields. Therefore, in domains such as external software design, "the 'distance' between the design concept and the final product is shorter than in, say, architecture" (p. 93). This does not imply, however, that in those domains design and implementation are not separated. It might, however, clarify results such as our observation that software designers find it particularly difficult to separate design from coding (Visser, 1987b).

---

[35] Cognitive-psychology research has shown differences in the processing of descriptions (thus, representations) of states and of processes.

Possible Types of Criteria and Forms of Evaluation

Domains differ in the means that may be used in order to evaluate design proposals (Malhotra et al., 1980, pp. 129–130). In engineering, "objective" measures of future artifacts' performance can be used. One can calculate whether a particular design (e.g., a bridge) meets functional requirements, for example, accommodation and maximum load. Using such measures, different proposals can be ranked somewhat objectively. The results of qualitative evaluation on subjective criteria, such as aesthetics, are much more difficult to translate into a "score," and thus to compare. Between these two extremes are evaluations based on different types of simulation, physical and mental.

Delay of Implementation

Something that has been considered to make the solving of social-science problems quite difficult is the "delay from the time a solution is proposed and accepted to when it is fully implemented. . . . Naturally, a good solution anticipates changes in conditions, but anticipation can be quite difficult" (Voss et al., 1983, p. 169). This remark that was formulated with respect to social-science problems is indeed particularly applicable in this domain, but may also hold for other design areas. The underlying factor may be of influence on the evaluation (through simulation and other means) of solution proposals.

Artifacts' Behavior Over Time

"Interactive systems are designed to have a certain behavior over time, whereas houses typically are not," according to Löwgren (1995, p. 94). Even if this assertion is questionable with respect to behavior in general, behavior over time is a dimension on which artifacts differ—and the types of behavior of different artifacts are quite diverse. An artifact's behavior over time may be related to its impact on people (discussed in the next subsection), to its use by people who are not necessarily transformed by this use, and to the interaction with people, as referred to by Löwgren. It may also be due to its deterioration.

Houses may not display "behavior" over time, but they change. Systems such as organizations or interactive systems are subject to specific types of change. "Good" designers anticipate the transformation—be it of an evolution or of a deterioration type—which the artifacts that they design may undergo. For social and interactive artifacts, this anticipation may be performed through simulation. The future behavior of certain technical artifacts may be anticipated based on calculations.

Impact of an Artifact on Human Activity and the Possibility to Anticipate It

Predicting people's future use of an artifact product and further anticipating the impact of the artifact on human activity, is one of the "characteristic and difficult properties" of designing (Carroll, 2000, p. 39). Indeed, "design has broad impacts on people. Design problems lead to transformations in the world that alter possibilities for human activity and experience, often in ways that transcend the boundaries of the original design reasoning" (Carroll, 2000, p. 21). Carroll et al. (1997) qualify this impact of certain systems on people as "transformative."

Even if all design has impact on people, certain domains seem more sensitive than others. An example of a domain in which design has particularly broad impacts on people is HCI: For example, the design of interactive systems enabling novel educational activities, that is, the domain with which Carroll et al. (1997, 1998) have been especially concerned. Yet, all design with social implications has impact on people.

The possibility of anticipating this impact may vary between domains, and not necessarily depending on the degree of impact. It depends, among others things, on the possibility of simulating the artifact, or testing it in other ways.

The Role of the User in the Design Process

In each domain of design, users are central—even if not always for the designers. Artifacts are to be used by people, even if this use can be more or less direct. Domains differ, however, with respect to the way in which they take into account the potential, future users and their use of the artifact. In design of HCI, for example, there has been much effort toward the integration of user data into the design. This integration has taken various forms. It may consist of designers—or other design participants who "know" the user, but are not the users themselves—introducing these data into the design. There are also approaches, such as participatory design, in which the users have themselves a voice in the design process (Carroll, 2006).

It seems likely that the number and variety of participants who take part in a design process influences this process, probably more its socio-organizational than its cognitive aspects. Yet, on a cognitive level, the difficulty of integration may augment with the number of different participants—and thus of representations—to be integrated. In addition, the participation of such "nontechnical" design participants as users are, may introduce a specific difficulty both for the users themselves and for their professional design "colleagues"—again not only of a cognitive nature (cf. the next subsection).

Individual Versus Collective Design

This dimension has been touched on earlier (see the section *Individual and Collective Design* in the *Introduction*), but not from the perspective adopted here. Certain artifacts are generally designed by an individual designer, while for others it is rare that they are designed by a designer working alone. Complexity and size of artifacts play a role, but are certainly not the only variables.

As noticed in the *Introduction*, we do not see any evidence to suppose that cooperation modifies the nature of the basic cognitive activities and operations implemented in design (i.e., generation, transformation, and evaluation of representations). Because cooperation proceeds through interaction, it introduces, however, specific activities and influences designers' representational structures (both on sociocognitive and emotional levels). We presented examples of specific activities, such as coordination, operative synchronization, construction of interdesigner compatible representations, conflict resolution, management of different representations through confrontation, articulation, and integration. We emphasized the important role of argumentative activities. Given our view of argumentation—activities aiming to modify the representations held by one's interlocutors—this seems evident.

The construction of interdesigner compatible representations, their existence beside designers' private representations, and their management also introduce factors that may add complexity to collective design situations compared to individual design.

Other Dimensions

In his paper "How Is a Piece of Software Like a Building? Toward General Design Theory and Methods" (presented at the 2003 NSF Workshop on Science of Design: Software Intensive Systems), Gross (2003) advances that pieces of software and buildings are alike on several dimensions. Most dimensions mentioned by Gross are not among those that we proposed previously: The size of software and buildings; their level of complexity; the type of use or user, which may change (more or less) or may remain constant; the degree to which components of the artifact may be subject to change or renewal (cf. *Artifacts' behavior over time*); their lifetime (which may be more or less extended; cf. *Artifacts' behavior over time*); the proportion of reusable components in the artifact's structure; the difference or equivalence between client and user; and the sanitary risks and safety concerns that particular uses or states of the artifact may introduce. According to Gross, on all these dimensions, pieces of software and buildings are alike (see also Kapor, 1996).

In the last subsection of this section, we discuss two fields of design that are often discussed as more or less, or even completely different from design in other application domains. These are software design and design of HCI. In fact, the resemblance between the two is also an object of discussion.

Software and HCI Design
Versus Other Forms of Design

In the cognitive design research literature, one frequently encounters allusions to, or implicit testimonies of the specific character of software design compared to other types of design—design of HCI is much less the object of discussion in this context. The responsible dimensions remain, however, unexplored.

In the preliminary analyses of their bibliographic cocitation analysis (presented earlier), Atwood et al. (2002) eliminated from further analysis "a set of authors representing Software Engineering design methodologies . . . [who] were found to be essentially unconnected with the remainder of the author set" (p. 129). "Software design has its own design literature" (p. 132). In the historical chapter in Part I (chap. 3), we also presented software-design studies as a specific domain (cf. also Détienne's, 2002a, book exclusively dedicated to design of software).

Indeed, general "design journals," such as *Design Studies*, *Design Issues*, or *The Journal of Design Research* rarely publish papers on software or HCI design. *Design Studies'* first paper (Thomas & Carroll, 1979/1984) was one of the three or four cognitively oriented studies on software-design activity that the journal has published in some 35 years. Two other, more recent, examples are Davies and Castell's (1992) analysis of empirical software-design studies and our own analysis of review meetings (D'Astous et al., 2004). If the papers that the journal accepts for publication are representative of the submissions, researchers submitting to *Design Studies* mainly come from architectural and urban design, engineering, and product design.

The separation between software and HCI, and other types of design, holds for scientific events as well. Conferences in the domain of cognitive design research concern software design and/or HCI (i.e., treated either together or singly), or other types of design.

When they announce "design" as their object (without further specification), conferences generally do not expect that research on software or HCI design will be presented. This holds for general conferences and workshops in the domain of design that are open to cognitive design studies.[36] There are also conferences that are not specifically dedicated to design, but open to papers on designing, such as conferences in the domain of ergonomics and human factors. These

---

[36] Examples of such conferences and workshops are ICED, the International Conference on Engineering Design; HBiD, the International Workshop on Human Behaviour in Designing; HI, the International Roundtable Conference on Computational and Cognitive Models of Creative Design (HI = Heron Island); DCC, the International Conference on Design Computing and Cognition; AID, the Artificial Intelligence in Design conference; VR, the Visual and Spatial Reasoning in Design conference; the International Design Conference DESIGN; and the Design Thinking Research Symposia (DTRS, the follow-up of the Research in Design Thinking symposia).

conferences are less devoted to a particular application domain.[37] Finally, there are conferences dedicated to collective or collaborative activities, often with a focus on computer support, in which researchers may present work on design from a cognitive viewpoint—often design of software systems.[38]

On the other hand, there are obviously conferences and workshops in the domain of software or HCI with, in principle, interest for cognitive approaches to design.[39] Conferences dedicated to HCI, however, often preferentially focus on technological solutions over analyses of cognitive activities (comparable to the CSCW conferences, see Note 38). CHI is a clear example; another one is IHM, the Francophone conference on human-machine interaction.

There are of course exceptions to these general affirmations. An example in the domain of conferences was the International Conference in honour of Herbert Simon, "The Sciences of Design. The Scientific Challenge for the 21st Century" (Lyon, France, 15–16 March 2002) (Forest et al., 2005; Perrin, 2002).

The potential specificity of HCI design. Design in the domain of HCI has been widely discussed for some 25 years. Some authors notice the similarities between the design of user interfaces and that of—at first sight—completely other types of artifacts, for example, architecture or automotive design. They base their judgment on the fact that in all these domains, "a good design relies on principles such as ease-of-use and providing functionality that meets real needs" (Ford & Marchak, 1997). Winograd (1996) considers that its user-oriented character makes software design comparable to architectural and graphic design, and different from engineering design. He also considers, however, that the design of interactive software is completely different from other software design (e.g., Winograd, 1997). Among all the arguments advanced for all these claims, none is based on cognitive analyses of the activities.

In the literature that seems to tackle design of HCI, designing as a cognitive *activity* is often "discovered" as an unexplored continent. Presenting some observations, evoking one or more questions, authors advance general ideas without any reference to the numerous results and models in the domain of cognitive

---

[37] Examples of such conferences are the European Conference on Cognitive Ergonomics, ECCE; the Ergonomics Society Annual Conference; and the conferences organized, once every 3 years, by the IEA, the International Ergonomics Association.

[38] Examples of such conferences are ECSCW, the European Conference on Computer-Supported Cooperative Work; Group, the International conference on supporting group work; and COOP, the International Conference on the Design of Cooperative systems. The CSCW conferences on Computer-Supported Cooperative Work focus more, either on technological solutions, or on ethnographical or ethnomethodological analyses of collective activities, which generally do not give much attention to underlying cognitive activities.

[39] Some examples of such conferences and workshops are the ESP Workshops on Empirical Studies of Programmers (until 1997); the PPIG Workshops organized by the Psychology of Programming Interest Group; CHI, the Annual SIGCHI Conference: Human Factors in Computing Systems; HCI, the British HCI Group Annual Conference; INTERACT, the IFIP TC13 International Conference on Human–Computer Interaction; and DIS, the conference on Designing Interactive Systems (see the conference page of ACM/SIGCHI, retrieved August 23, 2005, from http://www.acm.org/sigchi/conferences/).

design research. This approach may even be found among those involved in CHI, *the* conference in the domain of HCI. Both empirical work and modeling approaches of the cognitive aspects of design activity are rare among contributions to this conference. In 1995, CHI "instituted a new section called design briefings for presentation of notable designs and for discussion of how those designs came to be" (Winograd, 1996, p. xiv). These contributions did not discuss, however, the cognitive aspects of this "coming to be."

Another striking example is Dykstra-Erickson, Mackay, and Arnowitz's (2001) paper in the ACM (Association for Computing Machinery) bulletin of SIGCHI (Special Interest Community in Computer-Human Interaction). The authors present their article "as a *starting point* for [the] discussion of design [in SIGCHI]" (p. 109; the emphasis is ours). Seemingly unaware of the existing tradition in cognitive design studies, the "fundamental topic" of the paper is the question: "What is 'design,' and how do we define 'designers'?" (p. 109). In order to answer these questions, the authors introduce three different "perspectives on design": Norman's culturo-organisational view, Sarah Kuhn's sociotechnical approach, and Kelley's idea that "design" and "designer" have a much broader coverage than is generally acknowledged and that design, amongst other aspects, comprises visualizing. "A dilemma for the CHI community [is that] 'design' really isn't something that can be narrowly defined" (p. 113)—no cognitive perspective is referred to, however. The authors "propose" that design "needs to be further qualified" and that this could take place "at the CHI table" (p. 113).

Many studies in the domain of "interactive computing systems for human use" (as ACM SIGCHI defines HCI) seem to be related to designing. Nevertheless, with only a few exceptions, little attention is paid to the cognitive, dynamic aspects of the underlying activity (see, e.g., our analysis of user considerations guiding development and evaluation in Visser, 1987, and presented earlier)

Studies on design knowledge that is considered relevant for HCI designers rarely analyze *the way in which* designers *use* such knowledge (what we qualify here as the dynamic aspects of design; see also Visser, 2003). Referring to authorities such as Norman and Draper (1986), authors advance that "the model of design in HCI" emphasizes user-centeredness during the design process, but they do not describe *by which cognitive activities* designers of HCI systems may *realize* this user-centeredness in their design.

In theory, the design of HCI might exploit certain conclusions from software-design studies, which historically occupy a strong position in cognitive design research. However, many of these studies provide little data on actual design. As emphasized previously, most of them concern "programming" in the sense of coding.

The absence of studies in the domain of HCI design on cognitive, dynamic aspects of the activity may seem surprising. It may be due to the scarcity of cognitive psychologists in the HCI community. Ergonomists and other persons working in the domain of human factors, cognitively oriented or not, indeed often lack a research tradition into *activities* from a *cognitive* viewpoint—at least, at a microscopic level (Garrigou & Visser, 1998).

We wish to emphasize that this discussion is not intended to suggest that research in HCI is not occupied with relevant cognitive aspects. Research topics such as an artifact's impact on the user, its consequences for user involvement in design, and the possible ways to realize users' participation, are central to this area and have not been much examined in other domains of design.

Yet, at the end of this subsection *Software and HCI Design Versus Other Forms of Design*, the question remains: Is there specificity to software or to HCI design—and if so, what is it?

With respect to this entire section *Design as a Specific Cognitive Activity, In Spite of Design Taking Also Different "Forms,"* our list of candidate dimensions that might differentiate such forms of design is a start for their further elaboration and analysis. A next step would be to elucidate if indeed, and if so how, these and other differences influence design activity and its result, the artifact product. Still another step—announced in the introduction of this section—concerns examination of the fourth point of our hypothesis, namely that in spite of the possibly different forms of design, design remains a specific cognitive activity relative to other cognitive activities.

22.2. LINKING REPRESENTATIONAL STRUCTURES AND ACTIVITIES

Modeling involves abstracting and simplifying. Various criticisms that we have formulated in this book have concerned simplification. This holds not only for Simon and his SIP approach to design, but also for most early software-design models, and for Goel's (1995) model of design. In all three cases, we considered the artificially restricted conditions of data collection as an important explanatory factor. Another factor is the researcher's focus, as we noticed concerning Davies (1991a) and Ball and Ormerod (1995). From the start of our research in the domain of cognitive design studies on (see Table 6.1), our major object of modeling has been what we qualified as "actual design," thus, design as performed in a designer's usual working situation. Simplifying such design may be "necessary" for modeling, but it leads to neglect at least part of this design's specific characteristics that distinguish it from design as represented in normative models.

We may nevertheless formulate some *tendencies* that we have identified in our analysis of the different cognitive design studies presented in this book. They concern functional linkages between representational structures and activities.

We examine these linkages through the scheme "input representation - activity- output representation." Different dependencies exist: The input or output representation may constrain an activity (I_Repr or O_Repr -> Act), or be constrained by an activity (Act -> I_Repr or O_Repr), or two elements can mutually constrain each other (<->). The constraint may concern different aspects of an element. For representations, an example is their external or internal, notational or non-notational nature.

Our first observation is a case of Act -> I_Repr, Act constraining the notational nature of I_Repr. We suppose that activities concerned with gaining a new perspective on a representation (e.g., association, [re]interpretation, inductive inference) will more often be based on non-notational representations, such as sketches and other fluid forms of representation, than on notational representations, such as drafts with dimensions and tolerances.

The second case is one of Act -> O_Repr, where Act constrains the internal-external character of O_Repr. Interpretational activities will specifically result in internal representations—even if designers may of course use external representations in order to proceed to these activities.

The provisionality that non-notational forms of drawings may convey through roughness, and face-to-face, oral communication through specific phrasing and intonation, points to an Act -> I_Repr relation between non-notational representational systems and particular interactional activities in cooperative design.

The last two examples concern I_Repr -> Act. First, in domains of design concerned with physical artifacts, a common association exists between simulation-based evaluation and external representations (from two-dimensional drawings to three-dimensional mock-ups). Second, calculation and comparable activities will use notational, more formalized representational systems and, in design, they will be associated with particular notational types of representations such as drafts with dimensions and tolerances. These algorithmic activities may be used for the quantitative form of evaluation.

A next step to an integrated cognitive model of design will require identifying more such functional linkages and, especially, the organization of the different activities and corresponding representations into a structure, such as the blackboard framework that we have adopted in previous studies for the designers' organization of their activity (Visser, 1994a).

VI

Conclusion

In this conclusion, we first come back to two issues that have been central to this book: (a) SIP, SIT, and design and (b) design and problem solving. We establish a link between SIP- and SIT-inspired research, on the one hand, and studies into individual and collective design, on the other. Our design and problem-solving discussion strengthens our position that design involves problem solving, but is more than problem solving. We then conclude our book with a review and discussion of the original proposal that we have been introducing in this book: The characterization of design as a construction of representations, which opens up new possibilities for the study of design, and as a consequence provides new views on designers' needs for assistance and on the potential corresponding support modalities.

23 SIP and SIT, Individual and Collective Design

It was by reference to Simon's idea of human limited capacities leading to bounded rationality that Schön and Wiggins (1992) qualified designing as possessing "the conversational structure of seeing-moving-seeing" (p. 143). It may be interesting to notice that Simon himself did not realize the bounded-rationality view in his approach to design. As observed already, Simon did not apply to his general model of design the approach he adopted for social planning — and human economic behavior. He discussed the bounded-rationality idea, which was so central in his approach to economics (Simon, 1955), as relevant to his view of social planning. However, he did not apply it to his analysis of design in general, which he presented through discussions of two other domains, namely engineering and architectural design, without introducing notions related to cognitive limitations and their consequences for human activity.

As noticed several times in this book, our critique of the SIP approach to design concerns its over-systematic, and thereby impoverished, view of design. Simon indeed represents design, that is, engineering and architectural design, as much more orderly than it has been observed in studies on actual design activities — by the way, he never refers to such research. Simon's view applies to "simple," well-defined problems and to their processing, but does not represent the ill-defined problems that professional designers have to solve. This position has been substantiated through the discussion of six aspects on which the SIP approach tends to misrepresent design. Even if a model, by definition, simplifies its object, Simon's representation of design, in all its simplicity, does not render the specificity of design.

Greeno (1997, p. 5), once collaborating with Simon (Greeno & Simon, 1988), came to adopt an SIT perspective. Among the proponents of this family of approaches, we consider him as one of the most meticulous authors. We wish to recall the terms in which he presented SIP and SIT in the *Educational Researcher* debate on situated cognition/action. Greeno stated that the SIP perspective

> takes the theory of individual cognition as its basis and builds toward a broader theory by incrementally developing analyses of additional components that are considered as contexts. The situative perspective takes the theory of social and ecological interaction as its basis and builds toward a more comprehensive theory by developing increasingly detailed analyses of information structures in the contents of people's interactions. (p. 5)

This viewpoint is compatible with our view that the two positions are not contradictory, but complementary — not only because SIP would focus on individual design, and SIT on interaction in design.

As claimed in the confrontation of SIP and SIT approaches to design, we judge nevertheless that both Simon and Schön neglect essential aspects of design. Simon disregards the characteristics of the activity that lead to the "rich" nature of design (in particular, the role of problem-representation construction, the specificity of ill-defined problems solving, and the importance of other activities and operations than used in the classical transformation problems—cf. Simon's "nothing special" position). Schön gives detailed descriptions of extremely "rich" situations, but neither systematizes his observations nor makes the reflective step that we consider critical and necessary in order to attain a more abstract level on which more generally valid statements about design can be made. We saw, however, that current SIT-inspired research is evolving toward more explicit and operational theoretical frameworks. Yet, we concluded that the emphasis and focus of SIT-oriented researchers on situational resources neglect that it is the *designer* who, *using her or his knowledge and representational activities*, has to process the situation (identify and select elements in it, interpret it, transform it). As these processes depend on representational activities, we attribute such an essential role to the construction of representations— by the designer!

With respect to the introduction of the situation, and thus of other people, in the analysis of design, the switch of focus in cognitive design research from studies focusing on individual to studies focusing on collective design has entailed an evolution of the theoretical frameworks that researchers adopt as their reference. The purely cognitive framework based on the SIP approach is not sufficient for modeling individual design, but in order to address the collaborative, interactional nature of work, it is certainly inadequate. Because of its interest for situational resources, the SIT approach has *in principle* the potential to propose a more appropriate view on design. We have seen in this book how recent studies adopting the reflective-practitioner viewpoint are developing increasingly detailed analyses of central situativity notions—and not only in collective design settings.

24 Design Involves More Than Problem Solving

For somebody who is not familiar with the cognitive sciences, the term "problem solving" may seem inappropriate as applied to design. As we noted, however, the technical cognitive-psychology acceptation of "problem solving" covers a wide range of problem-centered activities that cover also those implemented in design. In formal terms, design may thus be qualified as problem solving.

Many authors state that design *is not* "problem solving." For Cross (2001a), designing is not "normal" problem solving (p. 81). Other authors consider that design is *not only* problem "solving," but also problem "setting," or "framing," to name just a few other activities advanced. Certain criticisms that SIT proponents address at the SIP approach are based, in our view, on terminological confusions, especially concerning "problem" and "problem solving." Yet, we cannot deny that many problem-centered activities covered by the term "problem solving" have not received much attention in the SIP research literature—which is the reference for problem-solving studies.

Another reaction is to consider that design *is no* "problem solving" such as this activity has been presented conventionally. In his paper first presented at the Workshop "User-Centred Design" during the International Conference in honour of Herbert Simon "The Sciences of Design. The Scientific Challenge for the 21st Century" (2002; published in 2006), Carroll (2006) describes how he

> first read "The Sciences of the Artificial" in the late 1970s [and] was intrigued by the idea that design could be considered a kind of problem solving; it stretched the concept of problem solving [he] had been taught as a graduate student in cognitive psychology. And [he] really liked the idea that design is not a residual category of problem solving, not a collection of miscellaneous loose ends, but rather lies right at the core of what humans do and what they are. (pp. 16–17)

This view might lead to open up the concept of problem solving. We adopt, however, still another position.

Even if design is problem solving in the sense that its requirements often will not evoke a memorized procedure, this does not inform us of the activities used in order to "solve" the corresponding "design problem," especially the generation and transformation of representations. Design thus involves *more than* problem solving. Characterizing design as "problem solving" does not capture its essence!

In this book, we defend that it is through different types of representation-construction activities—generation, transformation, and evaluation—which may take different forms (especially duplication, addition, detailing, concretization, modification, and revolutionizing) and refer to other activities and operations (such as analysis, [re]interpretation, association, confrontation, adjustment, integration, inference, restructuring, combining, and drawing).

Design is no transformation or structure-induction task, that is, the two types of problem-solving tasks for which classical cognitive-psychology studies have identified the basic operations and strategies. Even if "transformation" of representations is essential to design, the strategies and operations distinguished as typical for transformation problem-solving tasks (especially, systematic decomposition and hierarchical planning, and means-end analysis) are not at all central in the design activity.

25 Construction of Representations: Conclusion and Framework for Further Research

We have defined design as an activity that consists in specifying an artifact product, given requirements on that artifact. The requirements are global, abstract, and imprecise, thus incomplete; the final specifications are to be so detailed, precise, and concrete that they completely and explicitly specify the artifact product and its implementation. Both requirements and specifications are representations (both composite, generally). Therefore, the activity of design consists in the transformation of an input representation into an output representation, where these two representations are of a completely different nature, but represent the "same" artifact product. To be more precise, however, the input representation — the requirements — represents a great number of artifact products, whereas the output representation — the specifications — is indeed supposed to represent only one artifact product. The output representation instantiates one among the various possibilities provided by the abstract and global character of the input representation. Yet, in practice, given that only in theory an artifact's specifications are complete and entirely explicit, a design project's output representation still leaves space for different artifacts to be implemented. As noted, design continues during implementation.

Most features presented in this book as characteristic of design contribute to characterize the underlying activity as multifaceted: Its ill-definedness, complexity, ambiguity, the incomplete and especially the conflicting nature of its constraints — and the importance of representations, diverse with respect to their abstraction and precision, their internal and external, notational and non-notational character.

In an activity that functions by way of representations, knowledge plays a central role. Recognition — dependent on knowledge — is important as well: Simon was completely right in highlighting its role. Yet, we consider that Simon overestimated its importance relative to the controlled use of knowledge. In order to recognize a potentially relevant element of knowledge, there must be memory associations between target (features of the situation) and source (knowledge elements). Otherwise, controlled search or exploration has to be used, leading — if one is lucky — to unintended and unforeseen, but useful discoveries.

With regard to a professional activity as design, the knowledge on which both recognition and analogical reasoning, and other knowledge-intensive activities are based, is grounded mainly in professional experience (but see Visser, 1995b). Such knowledge also shapes the ill-definedness of a problem (but in the inverse sense, we suppose, as advanced by Simon, 1973/1984, p. 197, quoted earlier). For a designer with extensive and long experience in several application domains, a task that constitutes an ill-defined problem for an inexperienced colleague, may constitute a task with a more or less routine character. Knowledge-intensive tasks indeed evolve with respect to their "problem" and their "routine" character, their "ill-definedness" — and possibly even their "design" character.

Adopting an entirely "problem-solver oriented" approach, Thomas and Carroll (1979/1984, p. 222) propose viewing design, not as a particular "type" of problem, but as a "way of looking at" a problem — we might say a particular representation of a problem (cf. Simon's, 1969/1999b, focus-of-attention idea regarding changing one's representation). Theorem proving, for example, can be viewed as designing if, say, the requirement to stay inside certain formal rules is relaxed and creativity is allowed; in this case, although the initial goal was to prove a theorem, the goal in fact becomes "to find out something interesting." In the same way, "designing" a house by applying a set of standard rules to the stated requirements is no longer "design" from a cognitive viewpoint. Therefore, whether a task is "design" depends entirely on the person who faces the task. "Much of what [people] call technological progress may be viewed as a process of rendering ill-structured design problems as more well-structured procedures for accomplishing the same ends — without requiring design" (Thomas & Carroll, 1979/1984, p. 222). Indeed, if a "design task" is no longer open-ended, ill-defined, ambiguous, if its constraints are the object of agreement, a "design problem" can become a "transformation problem" — or even no longer constitute a "problem" at all!

The proposal formulated in this book is consequential in that it opens new directions, both for research and for development of support. Qualifying design as problem solving does not inform us of the activities and structures that designers are implementing. The analysis of design as construction of representations has led us to identify a number of activities and representational structures involved, and to propose a number of dimensions that may be of influence on them.

Clearly, this approach may also guide further research. Some examples of areas, which, in this text, have been discussed in more or less detail, or have only been alluded to, are the following:

- The shapes that the construction of representations may take, both on an individual and on a collective plane.
- The dimensions underlying forms of designing, and the differences they involve for both the activity and the representational structures.
- The representational structures constructed and used depending on these dimensions, and the role of representational formats in their construction and use.

A new vision of a domain provides not only new possibilities for research, but also new views of needs for assistance and of potential support modalities. If domain-specific but varied forms of knowledge, and representational activities and structures are effectively so essential in design, if designers' way of looking at their projects is indeed so critical to their success, this has implications for support systems. Given our observations, a valuable approach seems to be the development of systems — not only technological, but also methodological and organizational — that may prompt the development of such knowledge, for example, through designers' exploitation of their involvement in different types of projects (Falzon & Visser, 1989, formulated a comparable approach for analogical reasoning and its consequences for reuse) and through other reflective activities (Mollo & Falzon, 2004). The development of appropriate support modalities for particular representational activities and structures requires specific analysis. Such research may take advantage of the progress already obtained in the domain of software and HCI design (e.g., research on visualization and visual programming languages), from research on diagrammatic reasoning (see the diagrammatic reasoning site, retrieved 19 October 2005, from http:// www.hcrc.ed.ac.uk/gal/Diagrams/; see also Blackwell, 1997) and other research into representational formats and their exploitation (e.g., in research on multiple — external — representations; see Van Someren, Reimann, Boshuizen, & De Jong, 1988).

A final research focus that closes this book concerns educating designers in "practicing representation learning with and about representational forms," as analyzed in such an enlightening way by Greeno and Hall (1997). Indeed, if design is the construction of representations, an essential part of design education consists of learning to construct, interpret, and use representations. "This learning involves much more than simply learning to read and write symbols in arrangements corresponding to the accepted forms" (Greeno & Hall, 1997). Representations are — also — tools (cognitive artifacts) and, as holds for all tools, people do not "simply" "use" them: People contribute in their construction. Such construction needs to be learned, as needs to be learned how to fully take advantage of all its components in using the tool (not only existing notations, and conventions of interpretation, but also ways in which one may attribute alternative interpretations to a representation, and construct representations using nonstandard forms).

As tools and as input, output, intermediate, and intermediary objects, representations are the typical forms that cognitive artifacts may take in designing, that is, in "creating the artificial," as formulated by Simon (1969/1999b).[40]

---

[40] Compare the title of chapter 5 in Simon's *Sciences of the Artificial* (1969/1999b): "The Science of Design: Creating the Artificial."

References

Adams, R. S., Turns, J., & Atman, C. J. (2003). Educating effective engineering designers: The role of reflective practice. *Design Studies, 24*(3), 275–294.

Adelson, B. (1984). When novices surpass experts: The difficulty of a task may increase with expertise. *Journal of Experimental Psychology: Learning, Memory and Cognition, 10*(3), 483–495.

Adelson, B., Littman, D. C., Ehrlich, K., Black, J. B., & Soloway, E. (1985). Novice-expert differences in software design. In B. Shackel (Ed.), *INTERACT 84—1st IFIP International Conference on Human–Computer Interaction* (pp. 473–478). Amsterdam: North-Holland.

Adelson, B., & Soloway, E. (1988). A model of software design. In M. T. H. Chi, R. Glaser, & M. J. Farr (Eds.), *The nature of expertise* (pp. 185–208). Hillsdale, NJ: Lawrence Erlbaum Associates.

Agre, P. (1993). The symbolic worldview: Response to Vera and Simon. *Cognitive Science, 17*(1), 61–70.

Agre, P., & Chapman, D. (1987). Pengi: An implementation of a theory of activity. In *Proceedings of the Sixth National Conference on Artificial Intelligence* (pp. 268–272). Menlo Park, CA: American Association for Artificial Intelligence.

Akin, Ö. (1984). An exploration of the design process. In N. Cross (Ed.), *Developments in design methodology* (pp. 189–207). Chichester, England: Wiley. (Original work published 1979)

Akin, Ö. (1986a). A formalism for problem restructuring and resolution in design. *Environment and Planning B: Planning and Design, 13*, 223–232.

Akin, Ö. (1986b). *Psychology of architectural design*. London: Pion.

Akin, Ö. (1992). A structure and function based theory for design reasoning. In N. Cross, K. Dorst, & N. F. M. Roozenburg (Eds.), *Research in design thinking* (pp. 37–60). Delft, Netherlands: Delft University Press.

Akin, Ö. (2004). Strategic use of representation in architectural massing. *Design Studies, 25*, 31–40.

Akin, Ö., & Akin, C. (1997). On the process in creativity in puzzles, inventions and designs. *Automation in Construction, 7*, 123–138.

Alexander, C. (1984). The determination of components for an Indian village. In N. Cross (Ed.), *Developments in design methodology* (pp. 33–56). Chichester, England: Wiley. (Original work published 1963)

Alexander, I. (2001, September). *Book review: The reflective practitioner. How professionals think in action*. Retrieved October 12, 2005, from http://i.f.alexander.users.btopenworld.com/reviews/schon.htm

Anderson, J. R. (1976). *Language, memory, and thought*. Hillsdale, NJ: Lawrence Erlbaum Associates.

Anderson, J. R. (1978). Arguments concerning representations for mental imagery. *Psychological Review, 85*, 249–277.

Anderson, J. R. (1983). *The architecture of cognition*. Cambridge, MA: Harvard University Press.

Anderson, J. R. (1993). *Rules of the mind*. Hillsdale, NJ: Lawrence Erlbaum Associates.

Anderson, J. R., Reder, L. M., & Simon, H. A. (1996). Situated learning and education. *Educational Researcher, 25*(4), 5–11.

227

Anderson, J. R., Reder, L. M., & Simon, H. A. (1997). Rejoinder: Situated versus cognitive perspectives: Form versus substance. *Educational Researcher, 26*(1), 18–21.

Anzai, Y., & Simon, H. A. (1979). The theory of learning by doing. *Psychological Review, 86,* 124–140.

Araújo Carreira, M. H. (1997). *Modalisation linguistique en situation d'interlocution. Proxémique verbale et modalités en portugais* [Linguistic modalisation in interlocution situations. Verbal proxemics and modalities in Portuguese]. Louvain-Paris: Peeters.

Araújo Carreira, M. H. (2005). Politeness in Portugal: How to address others. In L. Hickey & M. Stewart (Eds.), *Politeness in Europe* (pp. 306–316). Clevedon, England: Multilingual Matters.

Archer, L. B. (1984). Systematic method for designers. In N. Cross (Ed.), *Developments in design methodology* (pp. 57–82). Chichester, England: Wiley. (Original work published 1965)

Asimov, M. (1962). *Introduction to design.* Englewood Cliffs, NJ: Prentice-Hall.

Atwood, M. E., Burns, B., Gairing, D., Girgensohn, A., Lee, A., Turner, T., et al. (1995). Facilitating communication in software development. In G. M. Olson & S. Schuon (Eds.), *Symposium on Designing interactive systems. Proceedings of the Conference on Designing Interactive Systems: Processes, practices, methods, and techniques (DIS'95)* (pp. 65–73). New York: ACM Press.

Atwood, M. E., Gross, M. D., McCain, K. W., & Williams, J. C. (2004, February). *Science of design: Why we need it and why it is so difficult to achieve.* Paper presented at the Human Computer Interaction Consortium Winter Workshop (HCIC'04), Winter Park, CO.

Atwood, M. E., McCain, K. W., & Williams, J. C. (2002, June). *How does the design community think about design?* Paper presented at the Conference on Designing Interactive Systems: Processes, practices, methods, and techniques (DIS'02), London.

@ctivités. revue électronique. (2004). Numéro spécial Activité et action/Cognition située [Activity and action/Situated cognition]. *@ctivités. revue électronique, 1*(2). Available at http://www.activites.org/

Baker, M., Détienne, F., Lund, K., & Séjourné, A. (in press). Analyse épistémique et argumentative de la conception collective en architecture. Etude des profils interactifs [Epistemic and argumentative analysis of collective architectural design. A study of interactive profiles]. In F. Détienne & V. Traverso (Eds.), *Méthodologies d'analyse de situations coopératives de conception. Le corpus MOSAIC* [Methodologies for analysing cooperative design situations. The MOSAIC corpus]. Nancy, France: Presses Universitaires de Nancy.

Ball, L. J., Evans, J. S. B. T., & Dennis, I. (1994). Cognitive processes in engineering design: A longitudinal study. *Ergonomics, 37*(11), 1753–1786.

Ball, L. J., & Ormerod, T. C. (1995). Structured and opportunistic processes in design: A critical discussion. *International Journal of Human–Computer Studies, 43,* 131–151.

Ball, L. J., & Ormerod, T. C. (2000). Putting ethnography to work: The case for a cognitive ethnography of design. *International Journal of Human–Computer Studies, 53,* 147–168.

Balzer, R. (1981). Transformation implementation: An example. *IEEE Transactions on Software Engineering, SE-7,* 3–14.

Bannon, L. J. (1995). The politics of design: Representing work. *Communications of the ACM, 38*(9), 66–68.

Bardram, J. (1997, September). *Plans as situated action: An activity theory approach to workflow systems.* Paper presented at the ECSCW'97 conference, Lancaster, England. Retrieved October 12, 2005, from http://www.daimi.au.dk/PB/525/PB-525.pdf

Barstow, D. R. (1984). A perspective on automatic programming. *AI Magazine, 5*(1), 5–27.

Basili, V. R., & Turner, A. J. (1975). Iterative enhancement: A practical technique for software development. *IEEE Transactions on Software Engineering, 1*(4), 390–396.

Bayazit, N. (2004). Investigating design: A review of forty years of design research. *Design Issues, 20*(1), 16–29.

Baykan, C. A. (1996). Design strategies. In N. Cross, H. Christiaans, & K. Dorst (Eds.), *Analysing design activity* (pp. 133–150). Chichester, England: Wiley.

Benyon, D. R. (1992). Task analysis and system design: The discipline of data. *Interacting with Computers, 4*(2), 246–259.

Berlin, L. M. (1993). Beyond program understanding: A look at programming expertise in industry. In C. R. Cook, J. C. Scholtz, & J. C. Spohrer (Eds.), *Empirical Studies of Programmers: Fifth workshop (ESP5)* (pp. 8–25). Norwood, NJ: Ablex.

Bertolotti, F., Macrì, D. M., & Tagliaventi, M. R. (2004). Social and organisational implications of CAD usage: A grounded theory in a fashion company. *New Technology, Work and Employment, 19*(2), 110–127.

Best, B. J., & Simon, H. A. (2000). Simulating human performance on the traveling salesman problem. In N. Taatgen & J. Aasman (Eds.), *Third International Conference on Cognitive Modeling* (pp. 42–49). Groningen, Netherlands: Universal Press.

Bhaskar, R., & Simon, H. A. (1977). Problem solving in semantically rich domains: An example from engineering thermodynamics. *Cognitive Science, 1*, 193–215.

Bibliography of Herbert A. Simon—1930–1950's. Retrieved October 12, 2005, from http://www.psy.cmu.edu/psy/faculty/hsimon/HSBib-1930-1950.html

Biggerstaff, T. J., & Perlis, A. J. (1989a). Introduction. In T. J. Biggerstaff & A. J. Perlis (Eds.), *Software reusability* (Vol. 1). New York: ACM Press.

Biggerstaff, T. J., & Perlis, A. J. (Eds.). (1989b). *Software reusability* (Vol. 2). Reading, MA: Addison-Wesley.

Biggerstaff, T. J., & Perlis, A. J. (Eds.). (1989c). *Software reusability: Concepts and models* (Vol. 1). New York: ACM Press.

Bisseret, A. (1990). Towards computer-aided text production. In P. Falzon (Ed.), *Cognitive ergonomics: Understanding, learning and designing human–computer interaction* (pp. 213–229). London: Academic Press.

Bisseret, A., Figeac-Letang, C., & Falzon, P. (1988). *Modeling opportunistic reasonings: The cognitive activity of traffic signal setting technicians* (Research Rep. No. 898). Rocquencourt, France: Institut National de Recherche en Informatique et en Automatique. Available at http://www.inria.fr/rrrt/rr-0898.html

Blackwell, A. F. (1997). *Diagrams about thoughts about thoughts about diagrams.* Retrieved October 19, 2005, from http://www.cl.cam.ac.uk/~afb21/publications/AAAI.html

Blessing, L. T. M. (1994). *A process-based approach to computer-supported engineering design.* Unpublished doctoral dissertation, Universiteit Twente, Enschede, Netherlands.

Blessing, L. T. M., Brassac, C., Darses, F., & Visser, W. (Eds.). (2000). *Analysing and modelling collective design activities. Proceedings of COOP 2000, Fourth International Conference on the Design of Cooperative Systems.* Rocquencourt, France: Institut National de Recherche en Informatique et en Automatique.

Bloch, H., Chemama, R., Gallo, A., Leconte, P., Le Ny, J.-F., Postel, J., et al. (Eds.). (1991). *Grand dictionnaire de la psychologie* [Encyclopedic dictionary of psychology]. Paris: Larousse.

Boehm, B. W. (1976). Software engineering. *IEEE Transactions on Computers, C-25*(12), 1226–1241.

Boehm, B. W. (1981). *Software engineering economics.* Englewood Cliffs, NJ: Prentice-Hall.

Bonnardel, N. (1989). *L'évaluation de solutions dans la résolution de problèmes de conception* [Solution evaluation in design problem solving] (Research Rep. No. 1072). Rocquencourt, France: Institut National de Recherche en Informatique et en Automatique.

Bonnardel, N. (1991a). Criteria used for evaluation of design solutions. In Y. Quéinnec & F. Daniellou (Eds.), *11th Congress of the International Ergonomics Association: Designing for everyone and everybody* (Vol. 2, pp. 1043–1045). London: Taylor & Francis.

Bonnardel, N. (1991b). L'évaluation de solutions dans la résolution de problèmes de conception et dans les systèmes experts critiques [Solution evaluation in design problem solving and in critic expert systems]. In D. Hérin-Aime, R. Dieng, J. P. Regouard, & J. P. Angoujiard (Eds.), *Knowledge modeling & expertise transfer* (pp. 371–381). Amsterdam: IOS Press.

Bonnardel, N. (1992). *Le rôle de l'évaluation dans les activités de conception* [The role of evaluation in design activities]. Unpublished doctoral dissertation, Université de Provence, Aix-en-Provence, France.

Boujut, J.-F., & Laureillard, P. (2002). A co-operation framework for product-process integration in engineering design. *Design Studies, 23*, 497–513.

Breuker, J., Wielinga, B., Van Someren, M. W., De Hoog, R., Schreiber, G., De Greef, P., et al. (1987). *Model-driven knowledge acquisition interpretation models* (Deliverable Task A1, Esprit Project 1098). Amsterdam: University of Amsterdam.

Brooks, R. A. (1991). Intelligence without representation. *Artificial Intelligence, 47*(1–3), 139–159.

Brooks, R. E. (1977). Towards a theory of the cognitive processes in computer programming. *International Journal of Man-Machine studies, 9*, 737–751.

Brown, D., & Chandrasekaran, B. (1989). *Design problem solving. Knowledge structures and control strategies*. London: Pitman.

Brown, J. S., Collins, A., & Duguid, P. (1989). Situated cognition and the culture of learning. *Educational Researcher, 18*(1), 32–42.

Bruner, J. S., Goodnow, J. J., & Austin, G. A. (1956). *A study of thinking*. New York: Wiley.

Bucciarelli, L. (1984). Reflective practice in engineering design. *Design Studies, 5*(3), 185–190.

Bucciarelli, L. (1988). An ethnographic perspective on engineering design. *Design Studies, 9*(3), 159–168.

Bucciarelli, L. (2002). Between thought and object in engineering design. *Design Studies, 23*(5), 219–231.

Buchanan, R. (1990, Octobre). *The "wicked problems" theory of design*. Paper presented at the Colloque Recherches sur le Design, Compiègne, France.

Buchanan, R. (1992). Wicked problems in design thinking. *Design Issues, 8*(2), 5–21. (Reprinted in *The idea of design*, V. Margolin, R. Buchanan, Eds., 1995, Cambridge, MA, MIT Press)

Buckingham Shum, S., & Hammond, N. (1994). Argumentation-based design rationale: What use at what cost? *International Journal of Human–Computer Studies, 40*, 603–652.

Burkhardt, J.-M. (1997). *Réutilisation de solutions en conception orientée-objet. Un modèle cognitif des mécanismes et représentations mentales* [Solution reuse in object-oriented design. A cognitive model of the mental mechanisms and representations] Unpublished doctoral dissertation, Université René Descartes de Paris V, Paris.

Burkhardt, J.-M., & Détienne, F. (1995). La réutilisation de solutions en conception de programmes informatiques [Solution reuse in computer-program design]. *Psychologie Française, 40*(1), 85–98.

Burns, C. M., & Vicente, K. J. (1995). A framework for describing and understanding interdisciplinary interactions in design. In G. M. Olson & S. Schuon (Eds.), *Symposium on Designing interactive systems. Proceedings of the Conference on Designing Interactive Systems: Processes, practices, methods, and techniques (DIS'95)* (pp. 97–103). New York: ACM Press.

Byrne, R. (1977). Planning meals: Problem-solving on a real data-base. *Cognition, 5*, 287–332.

Cacioppo, J. T., & Gardner, W. L. (1999). Emotion. *Annual Review of Psychology, 50*, 191–214.

Cagan, J., Kotovsky, K., & Simon, H. A. (2001). Scientific discovery and inventive engineering design: Cognitive and computational similarities. In E. K. Antonsson & J. Cagan (Eds.), *Formal engineering design synthesis* (pp. 442–465). Cambridge, England: Cambridge University Press.

Campbell, M., Cagan, J., & Kotovsky, K. (1999). A-Design: An agent-based approach to conceptual design in a dynamic environment. *Research in Engineering Design, 11*, 172–192.

Campbell, M., Cagan, J., & Kotovsky, K. (2000). Agent-based synthesis of electro-mechanical design configurations. *ASME Journal of Mechanical Design, 122*(1), 61–69.

Card, S., Moran, T. P., & Newell, A. (Eds.). (1983). *The psychology of human–computer interaction*. Hillsdale, NJ: Lawrence Erlbaum Associates.

Carlson, R. A. (1997). *Experienced cognition*. Mahwah, NJ: Lawrence Erlbaum Associates.

Carroll, J. M. (1995). *Scenario-based design. Envisioning work and technology in system development*. Hillsdale, NJ: Lawrence Erlbaum Associates.

Carroll, J. M. (2000). *Making use. Scenario-based design of human computer interactions.* Cambridge, MA: MIT Press.

Carroll, J. M. (2006). Dimensions of participation in Simon's design. *Design Issues, 22*(2), 3-18.

Carroll, J. M., & Rosson, M. B. (1985). Usability specifications as a tool in iterative development. In H. R. Hartson (Ed.), *Advances in human–computer interaction* (Vol. 1, pp. 1–28). Norwood, NJ: Ablex.

Carroll, J. M., Rosson, M. B., Chin, G., & Koenemann, J. (1997, August). *Requirements development: Stages of opportunity for collaborative needs discovery.* Paper presented at the Conference on Designing Interactive Systems: Processes, practices, methods, and techniques (DIS'97), Amsterdam.

Carroll, J. M., Rosson, M. B., Chin, G., & Koenemann, J. (1998). Requirements development in scenario-based design. *IEEE Transactions on Software Engineering, 24*(12), 1156–1170.

Carroll, J. M., Thomas, J. C., & Malhotra, A. (1980). Presentation and representation in design problem-solving. *British Journal of Psychology, 71,* 143–153.

Carroll, J. M., Thomas, J. C., Miller, L. A., & Friedman, H. P. (1980). Aspects of structure in design problem solving. *American Journal of Psychology, 93,* 269–284.

Chalmé, S., Visser, W., & Denis, M. (2000, September). *Cognitive aspects of urban route planning.* Proceedings of ICTTP 2000 (International Conference on Traffic and Transport Psychology) [CD-ROM], Bern, Switzerland.

Chalmé, S., Visser, W., & Denis, M. (2004). Cognitive effects of environmental knowledge on urban route planning strategies. In T. Rothengatter & R. D. Huguenin (Eds.), *Traffic and transport psychology. Theory and application* (pp. 61–71). Amsterdam: Elsevier.

Chambers, D., & Reisberg, D. (1985). Can mental images be ambiguous? *Journal of Experimental Psychology: Human Perception and Performance, 11,* 317–328.

Chatel, S., & Détienne, F. (1996). Strategies in object-oriented design. *Acta Psychologica, 91,* 245–269.

Chi, M. T. H., Glaser, R., & Farr, M. J. (Eds.). (1988). *The nature of expertise.* Hillsdale, NJ: Lawrence Erlbaum Associates.

Clancey, W. J. (1985). Heuristic classification. *Artificial Intelligence, 27,* 289–350.

Clancey, W. J. (1991). Situated cognition: Stepping out of representational flatland. *AI Communications—The European Journal on Artificial Intelligence, 4*(2–3), 109–112.

Clancey, W. J. (1993). Situated action: A neuropsychological interpretation. Response to Vera and Simon. *Cognitive Science, 17*(1), 87–116.

Clark, H. H., & Brennan, S. E. (1991). Grounding in communication. In L. Resnick, J.-M. Levine, & S. D. Teasley (Eds.), *Perspectives on socially shared cognition* (pp. 127–149). Washington, DC: American Psychological Association.

Conklin, J., & Begeman, M. L. (1988). gIBIS: A hypertext tool for exploratory policy discussion. *ACM Transactions on Office Information Systems, 6,* 303–331.

Cook, C. R., Scholtz, J. C., & Spohrer, J. C. (Eds.). (1993). *Empirical studies of programmers: Fifth Workshop (ESP5).* Norwood, NJ: Ablex.

Coyne, R. (2005). Wicked problems revisited. *Design Studies, 26*(1), 5–17.

Cross, N. (Ed.). (1984a). *Developments in design methodology.* New York: Wiley.

Cross, N. (1984b). Introduction to Part One: The management of design process. In N. Cross (Ed.), *Developments in design methodology* (pp. 1–7). Chichester, England: Wiley.

Cross, N. (2001a). Design cognition: Results from protocol and other empirical studies of design activity. In C. Eastman, W. M. McCracken, & W. C. Newstetter (Eds.), *Design knowing and learning: Cognition in design education* (pp. 79–103). Amsterdam: Elsevier.

Cross, N. (2001b). Strategic knowledge exercised by outstanding designers. In J. S. Gero & K. Hori (Eds.), *Strategic knowledge and concept formation III* (pp. 17–30). Sydney, Australia: University of Sydney, Key Centre of Design Computing and Cognition.

Cross, N. (2002). Creative cognition in design: Processes of exceptional designers. In T. Kavanagh & T. Hewett (Eds.), *Creativity and Cognition 2002* (pp. 14–19). New York: ACM Press.

Cross, N. (2004a). Expertise in design. Introduction. *The Journal of Design Research, 4*(2).

Cross, N. (2004b). Expertise in design: An overview. *Design Studies, 25*(5), 427–442.

Cross, N. (Ed.). (2004c). Expertise in design [Special issue]. *Design Studies, 25*(5).

Cross, N., Christiaans, H., & Dorst, K. (Eds.). (1996). *Analysing design activity*. Chichester, England: Wiley.

Cross, N., Dorst, K., & Roozenburg, N. F. M. (Eds.). (1992). *Research in design thinking*. Delft, Netherlands: Delft University Press.

Culverhouse, P. F. (1995). Constraining designers and their CAD tools. *Design Studies, 16*(1), 81–101.

Curtis, B. (1986). By the way, did anyone study any real programmers? In E. Soloway & S. Iyengar (Eds.), *Empirical Studies of Programmers: First Workshop (ESP1)* (pp. 256–262). Norwood, NJ: Ablex.

Damasio, A. R. (1994). *Descartes' error. Emotion, reason, and the human brain*. New York: Putnam.

Daniellou, F. (1999). *Le statut de la pratique et des connaissances dans l'intervention ergonomique de conception (Texte de l'habilitation à diriger des recherches présenté en 1992)* [The status of practice and knowledge in the ergonomic design intervention. Text of the Accreditation to supervise research presented in 1992]. Université Victor Segalen Bordeaux 2, Bordeaux, France.

Darke, J. (1984). The primary generator and the design process. In N. Cross (Ed.), *Developments in design methodology* (pp. 175–188). Chichester, England: Wiley. (Original work published 1979)

Darses, F. (1990a, September). *An assessment of the constraint satisfaction approach for design: A psychological investigation*. Paper presented at the Fifth European Conference on Cognitive Ergonomics (ECCE-5), Urbino, Italy.

Darses, F. (1990b). Constraints in design: Towards a methodology of psychological analysis based on AI formalisms. In D. Diaper, D. Gilmore, G. Gockton, & B. Shackel (Eds.), *INTERACT 90 — 3rd IFIP International Conference on Human–Computer Interaction* (pp. 135–139). Amsterdam: North Holland.

Darses, F. (1990c). *Gestion de contraintes au cours de la résolution d'un problème de conception de réseaux informatiques* [Constraint management during a computer-network design task] (Research Rep. No. 1164). Rocquencourt, France: Institut National de Recherche en Informatique et en Automatique.

Darses, F. (1991). The constraint satisfaction approach to design: A psychological investigation. *Acta Psychologica, 78*, 307–325.

Darses, F. (1994). *Gestion des contraintes dans la résolution des problèmes de conception* [Constraint management in design problem solving]. Unpublished doctoral dissertation, Université Paris 8, Saint-Denis, France.

Darses, F. (1997). L'ingénierie concourante. Un modèle en meilleure adéquation avec les processus cognitifs en conception [Concurrent engineering. A more adequate model to the cognitive processes in design]. In P. Brossard, C. Chanchevrier, & P. Leclair (Eds.), *Ingénierie concourante. De la technique au social* [Concurrent engineering. From the technical to the social] (pp. 39–55). Paris: Economica.

Darses, F. (2002, June). *A cognitive analysis of collective decision-making in the participatory design process*. Paper presented at the Seventh Participatory Design Conference (PDC'02), Malmö University, Malmö, Sweden.

Darses, F., Détienne, F., Falzon, P., & Visser, W. (2001). *COMET: A method for analysing collective design processes* (Research Rep. INRIA No. 4258). Rocquencourt, France: Institut National de Recherche en Informatique et en Automatique. Available at http://www.inria.fr/rrrt/rr-4258.html

Dasgupta, S. (1989). The structure of design processes. *Advances in Computers, 28*, 1–67.

Dasgupta, S. (2003). Multidisciplinary creativity: The case of Herbert A. Simon. *Cognitive Science, 27*(5), 683–708.

D'Astous, P., Détienne, F., Robillard, P. N., & Visser, W. (1998). Types of dialogs in evaluation meetings: An analysis of technical-review meetings in software development. In F. Darses & P. Zaraté (Eds.), *Third International Conference on the Design of Cooperative Systems (COOP'98)* (pp. 25–34). Rocquencourt, France: INRIA.

D'Astous, P., Détienne, F., Visser, W., & Robillard, P. N. (2004). Changing our view on design evaluation meetings methodology: A study of software technical review meetings. *Design Studies, 25*, 625–655.

D'Astous, P., Robillard, P. N., Détienne, F., & Visser, W. (2001). Quantitative measurements of the influence of participant roles during peer review meetings. *Empirical Software Engineering, 6*, 143–159.

Davies, S. P. (1991a). Characterizing the program design activity: Neither strictly top-down nor globally opportunistic. *Behaviour & Information Technology, 10*(3), 173–190.

Davies, S. P. (1991b). The role of notation and knowledge representation in the determination of programming strategy: A framework for integrating modes of programming behavior. *Cognitive Science, 15*, 547–572.

Davies, S. P., & Castell, A. M. (1992). Contextualizing design: Narratives and rationalization in empirical studies of software design. *Design Studies, 13*(4), 379–392.

De Terssac, G., & Chabaud, C. (1990). Référentiel opératif commun et fiabilité [Operative frames of reference and reliablity]. In J. Leplat & G. De Terssac (Eds.), *Les facteurs humains de la fiabilité dans les systèmes complexes* [Human factors of reliablity in complex systems] (pp. 111–139). Marseille: Octarès.

De Vries, E. (1994). *Structuring information for design problem solving* (Doctoral dissertation, Eindhoven University of Technology, Eindhoven, Netherlands). Den Haag, Netherlands: Koninklijke Bibliotheek.

Decortis, F., Leclercq, P., Boulanger, C., & Safin, S. (2004, March). *New digital environments to support creativity.* Paper presented at the International Seminar on Learning and Technology at Work, Institute of Education, London.

Demirbas, O. O., & Demirkan, H. (2003). Focus on architectural design process through learning styles. *Design Studies, 24*(5), 437–456.

Depraz, N., Varela, F., & Vermersch, P. (2003). *On becoming aware: A pragmatics of experiencing.* Amsterdam: John Benjamins.

Desmet, P. (2002). *Designing emotions.* Unpublished doctoral dissertation. Delft University of Technology, Delft, Netherlands.

Desmet, P., & Dijkhuis, E. (2003, June). *A wheelchair can be fun: A case of emotion-driven design.* Paper presented at the International Conference on Designing Pleasurable Products and Interfaces (DPPI-03), Pittsburgh, PA.

Desnoyers, L., & Daniellou, F. (1989). *SELF: The Francophone Ergonomics Society.* Retrieved October 14, 2005, from www.ergonomie-self.org/Pages/self/presentation/desnoyers.html

Détienne, F. (1990). Expert programming knowledge: A schema-based approach. In J.-M. Hoc, T. R. G. Green, R. Samurçay, & D. J. Gilmore (Eds.), *Psychology of programming* (pp. 205–222). London: Academic Press.

Détienne, F. (1991a). Reusing solutions in software design activity: An empirical study. *ACM SIGCHI Bulletin, 23*(4), 84–85.

Détienne, F. (1991b, August). *Solution reuse in expert design activity.* Paper presented at the International Conference on Cognitive Expertise, University of Aberdeen, United Kingdom.

Détienne, F. (1994). Constraints on design: Language, environment, code representation. In D. J. Gilmore, R. Winder, & F. Détienne (Eds.), *User-centred requirements for software engineering environments* (pp. 69–80). London: Springer.

Détienne, F. (2002a). *Software design. Cognitive aspects.* London: Springer.

Détienne, F. (2002b, June). *Supporting collaborative design: Current research issues.* Paper presented at the Fourteenth Annual Workshop of the Psychology of Programming Interest Group (PPIG14), London.

Détienne, F., & Burkhardt, J.-M. (2001). Des aspects d'ergonomie cognitive dans la réutilisation en génie logiciel [Cognitive-ergonomics aspects in software-engineering reuse]. *Techniques et Sciences Informatiques, 20*(4), 461–487.

Détienne, F., Burkhardt, J.-M., & Visser, W. (2004). Cognitive effort in collective software design: Methodological perspectives in cognitive ergonomics. In A. Jedlitschka & M. Ciolkowski (Eds.), *The future of empirical studies in software engineering. 2nd International Workshop, WSESE 2003* (pp. 23–31). Stuttgart, Germany: Frauenhofer IRB Verlag.

Détienne, F., & Falzon, P. (2001). Cognition and cooperation in design: The Eiffel research group. In M. Hirose (Ed.), *Human–Computer Interaction-Interact 2001* (pp. 879–880). Amsterdam: IOS Press.

Détienne, F., & Rist, R. S. (1995). Introduction to this special issue on Empirical studies of object-oriented design. *Human Computer Interaction, 10*(2–3), 121–128.

Détienne, F., & Traverso, V. (2003). Présentation des objectifs et du corpus analysé [Presentation of the objectives and the analyzed corpus]. In J. M. C. Bastien (Ed.), *Actes des Deuxièmes Journées d'Etude en Psychologie ergonomique-EPIQUE 2003* (pp. 217–221). Rocquencourt, France: Institut National de Recherche en Informatique et en Automatique.

Détienne, F., & Visser, W. (2006). Multimodality and parallelism in design interaction: Co-designers' alignment and coalitions. In P. Hassanaly, T. Herrmann, G. Kunau, & M. Zacklad (Eds.), *Cooperative systems design. Seamless integration of artifacts and conversations-Enhanced concepts of infrastructure for communication* (118–131). Amsterdam: IOS.

Détienne, F., Visser, W., D'Astous, P., & Robillard, P. N. (1999, May). *Two complementary approaches in the analysis of design team work: The functional and the interactional approach.* Position paper presented at the CHI99 Basic Research Symposium, Pittsburgh, PA.

Détienne, F., Visser, W., & Tabary, R. (in press). Articulation des dimensions graphico-gestuelle et verbale dans l'analyse de la conception collaborative [Articulating the graphico-gestural and verbal dimensions in the analysis of collaborative design]. *Psychologie de l'Interaction.*

Dillon, A., & Sweeney, M. (1988, September). *The application of cognitive psychology to CAD.* Paper presented at the HCI'88 Conference on People and Computers IV, Manchester, England.

Do, E. Y.-L., Gross, M. D., Neiman, B., & Zimring, C. M. (2000). Intentions in and relations among design drawings. *Design Studies, 21*(5), 483–503.

Donmoyer, R. (1996). Introduction. This issue: A focus on learning. *Educational Researcher, 25*(4), 4.

Dörner, D. (1999). Approaching design thinking research. *Design Studies, 20*(5), 407–416.

Dorst, K. (Ed.). (1995). Analysing design activity [Special issue]. *Design Studies, 16*(2).

Dorst, K. (1997). *Describing design. A comparison of paradigms.* Unpublished doctoral dissertation. Delft University of Technology, Delft, Netherlands.

Dorst, K., & Dijkhuis, J. (1995). Comparing paradigms for describing design activity. *Design Studies, 16*(2), 261–274.

Dreyfus, H. L. (2002). Intelligence without representation-Merleau-Ponty's critique of mental representation: The relevance of phenomenology to scientific explanation. *Phenomenology and the Cognitive Sciences, 1*(4), 367–383.

Dreyfus and Representationalism [Special issue]. (2002). *Phenomenology and the Cognitive Sciences, 1*(4).

Dunbar, K. (1999, July). *The cognitive paradox: Why reasoning in naturalistic settings is different from reasoning in the cognitive laboratory* [Abstract]. Plenary talk at the Second International Conference on Cognitive Science and the 16th Annual Meeting of the Japanese Cognitive Science Society Joint Conference (ICCS/JCSS99), International Conference Center, Waseda University, Tokyo.

Dwarakanath, S., & Blessing, L. T. M. (1996). Ingredients of the design process: A comparison between group and individual work. In N. Cross, H. Christiaans, & K. Dorst (Eds.), *Analysing design activity* (pp. 93–116). Chichester, England: Wiley.

Dykstra-Erickson, E., Mackay, W., & Arnowitz, J. (2001). Perspectives: Trialogue on design (of). *Interactions, 8*(2), 109–117.

Dym, C. L., & Little, P. (1999). *Engineering design: A project-based introduction.* New York: Wiley.

Eastman, C. (1969). Cognitive processes and ill-defined problems: A case study of design. In D. Walker & L. M. Norton (Eds.), *IJCAI'69, International Joint Conference on Artificial Intelligence* (pp. 669–690). San Mateo, CA: Kaufmann.

Eastman, C. (1970). On the analysis of intuitive design processes. In G. T. Moore (Ed.), *Emerging methods in environmental design and planning. Proceedings of the First International Design Methods Conference* (pp. 21–37). Cambridge, MA: MIT Press.

Eastman, C. (1999, April). *Representation of design processes.* Invited keynote address at the Conference on Design Thinking, MIT, Boston, MA. Retrieved March 8, 2006, from http://www.coa.gatech.edu/phd/research/references/design-thinking.pdf

Eastman, C. (2001). New directions in design cognition: Studies of representation and recall. In C. Eastman, W. M. McCracken, & W. C. Newstetter (Eds.), *Design knowing and learning: Cognition in design education* (pp. 147–198). Amsterdam: Elsevier.

Eckersley, M. (1988). The form of design processes: A protocol analysis study. *Design Studies, 9*(2), 86–94.

Eckert, C., & Stacey, M. (2000). Sources of inspiration: A language of design. *Design Studies, 21*(5), 523–538.

Edmonds, E., Riecken, R. D., Satherley, R., Stennin, K., & Visser, W. (1994). Computers and creative thought. In A. Cohn (Ed.), *Proceedings of ECAI 94. 11th European Conference on Artificial Intelligence* (pp. 779–784). Chichester, England: Wiley.

Ericsson, K. A., & Simon, H. A. (1980). Verbal reports as data. *Psychological Review, 87,* 215–251.

Ericsson, K. A., & Simon, H. A. (1993). *Protocol analysis. Verbal reports as data* (rev. ed.). Cambridge, MA: MIT Press. (Original work published 1984)

Expertise in Design [Special issue]. (2004). *The Journal of Design Research, 4*(2).

Fagan, M. E. (1976). Design and code inspections to reduce errors in program development. *IBM Systems Journal, 15*(3), 182–211.

Falzon, P. (1994). Dialogues fonctionnels et activité collective [Functional dialogs and collective activity]. *Le Travail Humain, 57*(4), 299–312.

Falzon, P. (2004). Préface [Preface]. In P. Falzon (Ed.), *Ergonomie* [Ergonomics] (pp. 11–13). Paris: Presses Universitaires de France.

Falzon, P., Darses, F., & Béguin, P. (1996, June). *Collective design processes.* Paper presented at COOP'96, Second International Conference on the Design of Cooperative Systems, Juan les Pins, France.

Falzon, P., & Visser, W. (1989). Variations in expertise: Implications for the design of assistance systems. In G. Salvendy & M. J. Smith (Eds.), *Designing and using human–computer interfaces and knowledge based systems* (Vol. 2, pp. 121–128). Amsterdam: Elsevier.

Feitelson, J., & Stefik, M. J. (1977). *A case study of the reasoning in a genetics experiment* (HPP Rep. No. HPP-77-18). Stanford, CA: Stanford University, Computer Science Department, Heuristic Programming Project.

Fischer, G. (2000). Symmetry of ignorance, social creativity, and meta-design [Internet version: Social creativity: Bringing different points of view together]. *Knowledge-Based Systems, 13*(7–8), 527–537. Retrieved October 12, 2005, from http://l2003d.cs.colorado.edu/~gerhard/papers/kbs2000.pdf

Fischer, G., Lemke, A. C., McCall, R., & Morch, A. I. (1991). Making argumentation serve design. *Human-Computer Interaction, 6*(3–4), 393–419.

Fischer, G., Nakakoji, K., Ostwald, J., Stahl, G., & Sumner, T. (1998). Embedding critics in design environments. In M. Maybury & W. Wahlster (Eds.), *Readings in intelligent user interfaces* (pp. 537–561). San Francisco: Morgan Kaufman.

Ford, S., & Marchak, F. M. (1997). The future of visual interaction design? *ACM SIGCHI Bulletin, 29*(1), 10–11.

Forest, J., Méhier, C., & Micaëlli, J.-P. (Eds.). (2005). *Pour une science de la conception. Fondements, méthodes, pratiques* [For a design science. Foundations, methods, practices]. Belfort-Montbéliard, France: Université de Technologie de Belfort-Montbéliard.

Foz, A. (1973). Observations on designer behavior in the parti. *DMG-DRS Journal: Design Research and Methods, 7*(4), 320–323.

Frankenberger, E., & Badke-Schaub, P. (1999). Special issue on Empirical studies of engineering design in Germany. Editorial. *Design Studies, 20*(5), 397–400.

French, M. J. (1971). *Engineering design: The conceptual stage.* London: Heinemann.

Fricke, G. (1992). Experimental investigation of individual processes in engineering (part 2). In N. Cross, K. Dorst, & N. F. M. Roozenburg (Eds.), *Research in design thinking* (pp. 105–109). Delft, Netherlands: Delft University Press.

Fricke, G. (1999). Successful approaches in dealing with differently precise design problems. *Design Studies, 20*(5), 417–430.

Friedman, K. (2003). Theory construction in design research: Criteria, approaches, and methods. *Design Studies, 24,* 507–522.

Galle, P. (1996). Replication protocol analysis: A method for the study of real world design thinking. *Design Studies, 17*(2), 181–200.

Garrigou, A., & Visser, W. (1998, February). *L'articulation d'approches macroscopiques et microscopique en ergonomie: une tentative de prise de recul sur une pratique de recherche en cours* [Articulating macroscopic and microscopic approaches in ergonomics: An attempt to stand back from a research practice in progress]. Paper presented at the Deuxièmes Journées "Recherche et Ergonomie" de la SELF (Société d'Ergonomie de Langue Française), Toulouse, France.

Gentner, D., & Stevens, A. L. (Eds.). (1983). *Mental models.* Hillsdale, NJ: Lawrence Erlbaum Associates.

Gero, J. S. (1990). Design prototypes: A knowledge representation schema for design. *AI Magazine, 11*(4), 26–36.

Gero, J. S. (Ed.). (1991). *AID'91, Artificial Intelligence in Design'91.* Oxford, England: Butterworth-Heinemann.

Gero, J. S. (Ed.). (1992). *AID'92, Artificial Intelligence in Design'92.* Boston: Kluwer.

Gero, J. S. (1998a). Conceptual designing as a sequence of situated acts. In I. Smith (Ed.), *Artificial intelligence in structured engineering* (pp. 165–177). Berlin, Germany: Springer.

Gero, J. S. (1998b, July). *Emergence in designing.* Paper presented at the "Emergence in design" Workshop held in conjunction with AID'98, the 5th International Conference on Artificial Intelligence in Design'98, Lisbon, Portugal.

Gero, J. S. (1998c). Towards a model of designing which includes its situatedness. In H. Grabowski, S. Rude, & G. Grein (Eds.), *Universal design theory* (pp. 47–56). Aachen, Germany: Shaker Verlag.

Gero, J. S. (1999a). Constructive memory in design thinking. In G. Goldschmidt & W. Porter (Eds.), *Design Thinking Research Symposium: Design representation* (pp. 29–35). Cambridge, MA: MIT Press.

Gero, J. S. (1999b). Recent design science research: Constructive memory in design thinking. *Architectural Science Review, 42,* 3–5.

Gero, J. S. (2002). Computational models of creative designing based on situated cognition. In T. Kavanagh & T. Hewett (Eds.), *Creativity and Cognition 2002* (pp. 3–10). New York: ACM Press.

Gero, J. S., & Kannengiesser, U. (2004). The situated function–behaviour–structure framework. *Design Studies, 25,* 373–391.

Gero, J. S., & McNeill, T. (1998). An approach to the analysis of design protocols. *Design Studies*, *19*(1), 21–61.

Gero, J. S., & Tang, H.-H. (2001). The differences between retrospective and concurrent protocols in revealing the process-oriented aspects of the design process. *Design Studies*, *22*(3), 283–295.

Gick, M. L., & Holyoak, K. J. (1980). Analogical problem solving. *Cognitive Psychology*, *12*, 306–355.

Gick, M. L., & Holyoak, K. J. (1983). Schema induction and analogical transfer. *Cognitive Psychology*, *15*, 1–38.

Gilhooly, K. J. (1989). Human and machine problem solving: Toward a comparative cognitive science. In K. J. Gilhooly (Ed.), *Human and machine problem solving* (pp. 1–12). New York: Plenum.

Gilmore, D. J. (1990a). Expert programming knowledge: A strategic approach. In J.-M. Hoc, T. R. G. Green, R. Samurçay, & D. J. Gilmore (Eds.), *Psychology of programming* (pp. 223–234). London: Academic Press.

Gilmore, D. J. (1990b). Methodological issues in the study of programming. In J. M. Hoc, T. R. G. Green, R. Samurçay, & D. J. Gilmore (Eds.), *Psychology of programming* (pp. 83–98). London: Academic Press.

Gilmore, D. J., Winder, R., & Détienne, F. (Eds.). (1994). *User-centred requirements for software engineering environments*. Berlin, Germany: Springer.

Glaser, R. (1986). On the nature of expertise. In F. Klix & H. Hagendorff (Eds.), *Human memory and cognitive performances*. Amsterdam: North-Holland.

Glaser, R., & Chi, M. T. H. (1988). Overview. In M. T. H. Chi, R. Glaser, & M. J. Farr (Eds.), *The nature of expertise* (pp. xv–xxviii). Hillsdale, NJ: Lawrence Erlbaum Associates.

Glasgow, J., Narayanan, N. H., & Chandrasekaran, B. (Eds.). (1995). *Diagrammatic reasoning: Cognitive and computational perspectives*. Cambridge, MA: MIT Press.

Goel, V. (1994). A comparison of design and nondesign problem spaces. *Artificial Intelligence in Engineering*, *9*, 53–72.

Goel, V. (1995). *Sketches of thought*. Cambridge, MA: MIT Press.

Goel, V. (1999). Cognitive role of ill-structured representations in preliminary design. In J. S. Gero & B. Tversky (Eds.), Visual and spatial reasoning in design (pp. 131–143). Sydney, Australia: University of Sydney, Key Centre of Design Computing and Cognition. Retrieved March 30, 2006, from http://www.arch.usyd.edu.au/kcdc/books/VR99/goel.html

Goel, V., & Pirolli, P. (1989). Motivating the notion of generic design within information-processing theory: The design problem space. *AI Magazine*, *10*(1), 18–36.

Goel, V., & Pirolli, P. (1992). The structure of design problem spaces. *Cognitive Science*, *16*, 395–429.

Goldschmidt, G. (1991). The dialectics of sketching. *Creativity Research Journal*, *4*(2), 123–143.

Goldschmidt, G. (1995). The designer as a team of one. *Design Studies*, *16*(2), 189–209.

Gray, W. D., & Boehm-Davis, D. A. (Eds.). (1996). *Empirical Studies of Programmers: Sixth Workshop (ESP6)*. New Brunswick, NJ: Ablex.

Greco, A. (1995a). The concept of representation in psychology. *Cognitive Systems*, *4*(2), 247–255.

Greco, A. (1995b). Introduction [to the Special issue on Representation]. *Cognitive Systems*, *4*(2), 119–129.

Green, T. R. G. (1980). Programming as a cognitive activity. In H. T. Smith & T. R. G. Green (Eds.), *Human interaction with computers* (pp. 271–320). London: Academic Press.

Green, T. R. G. (1990). Programming languages as information structures. In J.-M. Hoc, T. R. G. Green, R. Samurçay, & D. J. Gilmore (Eds.), *Psychology of programming* (pp. 117–137). London: Academic Press.

Greeno, J. G. (1978). Natures of problem-solving abilities. In W. K. Estes (Ed.), *Handbook of learning and cognitive processes* (Vol. 5, pp. 239–270). Hillsdale, NJ: Lawrence Erlbaum Associates.

Greeno, J. G. (1997). Response: On claims that answer the wrong questions. *Educational Researcher*, *26*(1), 5–17.

Greeno, J. G. (1998). The situativity of knowing, learning, and research. *American Psychologist, 53*(1), 5–26.

Greeno, J. G., & Hall, R. P. (1997). *Practicing representation learning.* Retrieved October 28, 2005, from http://www.pdkintl.org/kappan/k_v78/k9701gre.htm

Greeno, J. G., & Moore, J. L. (1993). Situativity and symbols: Response to Vera and Simon. *Cognitive Science, 17*(1), 49–60.

Greeno, J. G., & Simon, H. A. (1988). Problem solving and reasoning. In R. C. Atkinson, R. J. Herrnstein, G. Lindzey, & R. D. Luce (Eds.), *Stevens' handbook of experimental psychology* (2nd ed., Vol. 2, pp. 589–672). New York: Wiley.

Gross, M. D. (1996). The Electronic Cocktail Napkin. Computer support for working with diagrams. *Design Studies, 17*(1), 53–69.

Gross, M. D. (2003, November). *How is a piece of software like a building? Toward general design theory and methods.* Paper presented at the National Science Foundation (NSF) Invitational Workshop on Science of Design: Software Intensive Systems, Airlie Center, VA.

Guindon, R. (1990a). Designing the design process: Exploiting opportunistic thoughts. *Human Computer Interaction, 5*, 305–344.

Guindon, R. (1990b). Knowledge exploited by experts during software system design. *International Journal of Man-Machine Studies, 33*, 279–304.

Guindon, R., Krasner, H., & Curtis, B. (1987). Breakdowns and processes during the early activities of software design by professionals. In G. M. Olson, S. Sheppard, & E. Soloway (Eds.), *Empirical Studies of Programmers: Second Workshop (ESP2).* Norwood, NJ: Ablex.

Günther, J., Frankenberger, E., & Auer, P. (1996). Investigation of individual and team design processes. In N. Cross, H. Christiaans, & K. Dorst (Eds.), *Analysing design activity* (pp. 117–132). Chichester, England: Wiley.

Hamel, R. (1989). Design process and design problems in architecture. *Journal of Environmental Psychology, 9*, 73–77.

Hamel, R. (1995). *Psychology and design research.* Retrieved October 12, 2005, from http://www.designresearch.nl/PDF/DRN1995_Hamel.pdf

Hayes, J. R., & Flower, L. S. (1980). Identifying the organization of writing processes. In L. W. Gregg & E. R. Steinberg (Eds.), *Cognitive processes in writing* (pp. 3–30). Hillsdale, NJ: Lawrence Erlbaum Associates.

Hayes-Roth, B., & Hayes-Roth, F. (1979). A cognitive model of planning. *Cognitive Science, 3*, 275–310.

Hayes-Roth, B., Hayes-Roth, F., Rosenschein, S., & Cammarata, S. (1979, August). *Modeling planning as an incremental, opportunistic process.* Paper presented at the 6th International Joint Conference on Artificial Intelligence, Tokyo.

Herbsleb, J. D., Klein, H., Olson, G. M., Brunner, H., Olson, J. S., & Harding, J. (1995). Object-oriented analysis and design in software project teams. *Human–Computer Interaction, 10*(2–3), 249–292.

Heylighen, A., & Verstijnen, I. M. (2003). Close encounters of the architectural kind. *Design Studies, 24*(4), 313–326.

Heylighen, F. (1988). Formulating the problem of problem-formulation. In R. Trappl (Ed.), *Cybernetics and systems* (pp. 949–957). Dordrecht, Netherlands: Kluwer.

Hoc, J.-M. (1988a). *Cognitive psychology of planning.* London: Academic Press.

Hoc, J.-M. (1988b). Towards effective computer aids to planning in computer programming. Theoretical concern and empirical evidence drawn from assessment of a prototype. In G. C. van der Veer, T. R. G. Green, J.-M. Hoc, & D. Murray (Eds.), *Working with computers: Theory versus outcomes* (pp. 215–247). London: Academic Press.

Hoc, J.-M., & Carlier, X. (2002). Role of a common frame of reference in cognitive cooperation: Sharing tasks between agents in air traffic control. *Cognition, Work & Technology, 4*, 37–47.

Hoc, J.-M., Green, T. R. G., Samurçay, R., & Gilmore, D. J. (Eds.). (1990). *Psychology of programming.* London: Academic Press.

Hollan, J., Hutchins, E., & Kirsh, D. (2000). Distributed cognition: Toward a new foundation for human–computer interaction research. *ACM Transactions on Computer-Human interaction, 7*(2), 174–196. Reprinted in *Human–computer interaction in the new millennium*, pp. 75–94, by J. M. Carroll, 2000, Reading, MA: Addison-Wesley.

Hollnagel, E. (2002, March). *From human-centred to function-centred design.* Paper presented at the Workshop "User-Centred Design" at the International Conference in honour of Herbert Simon, "The Sciences of Design. The Scientific Challenge for the 21st Century," INSA, Lyon, France.

Holyoak, K. J. (1984). Analogical thinking and human intelligence. In R. J. Sternberg (Ed.), *Advances in the psychology of human intelligence* (Vol. 5, pp. 199–230). Hillsdale, NJ: Lawrence Erlbaum Associates.

Hsu, W., & Liu, B. (2000). Conceptual design: Issues and challenges. *Computer-Aided Design, 32*(14), 849–850.

Hubka, V., & Eder, W. E. (1987). A scientific approach to engineering design. *Design Studies, 8*(3), 123–137.

Hutchins, E. (1995). How a cockpit remembers its speed. *Cognitive Science, 19*(3), 265–288.

Hutchins, E. (1996). *Cognition in the wild.* Cambridge, MA: MIT Press.

Hwang, T. S., & Ullman, D. G. (1990). The design capture system: Capturing back-of-the-envelope sketches. *Journal of Engineering Design, 1*(4), 339–353.

Jackson, M. A. (1975). *Principles of program design.* New York: Academic Press.

Jackson, M. A. (1983). *System development.* Englewood Cliffs, NJ: Prentice-Hall.

Jedlitschka, A., & Ciolkowski, M. (Eds.). (2004). *The future of empirical studies in software engineering. 2nd International Workshop, WSESE 2003.* Stuttgart, Germany: Frauenhofer IRB Verlag.

Jeffries, R., Turner, A. A., Polson, P. G., & Atwood, M. E. (1981). The processes involved in designing software. In J. R. Anderson (Ed.), *Cognitive skills and their acquisition* (pp. 255–283). Hillsdale, NJ: Lawrence Erlbaum Associates.

Johnson-Laird, P. N. (1983). *Mental models. Towards a cognitive science of language, inference, and consciousness.* Cambridge, England: Cambridge University Press.

Johnson-Laird, P. N. (1989). Analogy and the exercise of creativity. In S. Vosniadou & A. Ortony (Eds.), *Similarity and analogical reasoning.* Cambridge, England: Cambridge University Press.

Jones, J. C. (1984). A method of systematic design. In N. Cross (Ed.), *Developments in design methodology.* Chichester, England: Wiley. (Original work published 1963)

Jones, T. C. (1984). Reusability in programming: A survey of the state of the art. *IEEE Transactions on Software Engineering, SE-10*(5), 488–493.

Kant, E. (1985). Understanding and automating algorithm design. *IEEE Transactions on Software Engineering, SE-11*, 1361–1374.

Kant, E., & Newell, A. (1984). Problem solving techniques for the design of algorithms. *Information Processing and Management, 20*(1–2), 97–118.

Kaplan, C. A., & Simon, H. A. (1990). In search of insight. *Cognitive Psychology, 22*, 374–419.

Kapor, M. (1996). A software design manifesto. In T. Winograd (Ed.), *Bringing design to software* (pp. 1–9). New York: ACM Press.

Karsenty, L. (1991a). Design strategies in database conceptual modelling. In Y. Quéinnec & F. Daniellou (Eds.), *11th Congress of the International Ergonomics Association: Designing for everyone and everybody.* London: Taylor & Francis.

Karsenty, L. (1991b). *Le dialogue de validation d'un schéma conceptuel des données* [Dialogs used to validate conceptual data schemata] (Research Rep. No. 1551). Rocquencourt, France: Institut National de Recherche en Informatique et en Automatique.

Kavakli, M., & Gero, J. S. (2001). Sketching as mental imagery processing. *Design Studies, 22*(4), 347–364.

Kavakli, M., Scrivener, S. A. R., & Ball, L. J. (1998). Structure in idea sketching behaviour. *Design Studies, 19*(4), 485–517.

Keane, M. T. (1994). Analogical asides on case-based reasoning. In S. Wess, K.-D. Althoff, & M. M. Richter (Eds.), *Topics in case-based reasoning, First European Workshop, EWCBR-93* (pp. 21–32). Amsterdam: Springer.

Kelley, D., & Hartfield, B. (1996). The designer's stance. In T. Winograd (Ed.), *Bringing design to software* (pp. 151–170). New York: ACM Press.

Kerbrat-Orecchioni, C. (1990). *Les interactions verbales* [Verbal interactions] (Vol. 1). Paris: Armand Colin.

Kerbrat-Orecchioni, C. (1992). *Les interactions verbales* [Verbal interactions] (Vol. 2). Paris: Armand Colin.

Kerbrat-Orecchioni, C. (1994). *Les interactions verbales* [Verbal interactions] (Vol. 3). Paris: Armand Colin.

Kim, J., Javier-Lerch, F., & Simon, H. A. (1995). Internal representation and rule development in object-oriented design. *ACM Transactions on Computer-Human Interaction, 2*(4), 357–390.

Kitchenham, B., & Carn, R. (1990). Research and practice: Software design methods and tools. In J.-M. Hoc, T. R. G. Green, R. Samurçay, & D. J. Gilmore (Eds.), *Psychology of programming* (pp. 271–284). London: Academic Press.

Klahr, D., & Simon, H. A. (2001). What have psychologists (and others) discovered about the process of scientific discovery? *Current Directions in Psychological Science, June, 10*(3), 75–79.

Klein, M., & Lu, S. (1989). Conflict resolution in cooperative design. *Artificial Intelligence in Engineering, 4*, 168–180.

Koenemann-Belliveau, J., Moher, T. G., & Robertson, S. P. (Eds.). (1991). *Empirical Studies of Programmers: Fourth Workshop (ESP4)*. New Brunswick, NJ: Ablex.

Kovordányi, R. (1999). *Modeling and simulating inhibitory mechanisms in mental image reinterpretation. Towards cooperative human–computer creativity.* Linköping Studies in Science and Technology, Dissertation No. 589, Linköpings University, Linköping, Sweden.

Krasner, H., Curtis, B., & Iscoe, N. (1987). Communication breakdowns and boundary spanning activities on large programming projects. In G. M. Olson, S. Sheppard, & E. Soloway (Eds.), *Empirical Studies of Programmers: Second Workshop (ESP2)* (pp. 47–64). Norwood, NJ: Ablex.

Krauss, R. I., & Myer, R. M. (1970). Design: A case history. In G. T. Moore (Ed.), *Emerging methods in environmental design and planning* (pp. 11–20). Cambridge, MA: MIT Press.

Kulkarni, D., & Simon, H. A. (1988). The processes of scientific discovery: The strategy of experimentation. *Cognitive Science, 12*, 139–175.

Kunz, W., & Rittel, H. W. J. (1970). *Issues as elements* (Working Paper No. 131; reprinted 1979). Retrieved August 5, 2005, from http://www-iurd.ced.berkeley.edu/pub/abstract_wp131.htm

Kyng, M., & Greenbaum, J. (1991). *Design at work*. Hillsdale, NJ: Lawrence Erlbaum Associates.

Lange, B. M., & Moher, T. G. (1989). Some strategies for reuse in an object-oriented programming environment. In K. Bice & C. Lewis (Eds.), *Proceedings of CHI'89, Conference on Human Factors in Computing Systems: Wings for the mind* (pp. 69–73). New York: ACM Press.

Langley, P., Simon, H. A., Bradshaw, G. L., & Zytkow, J. M. (1987). *Scientific discovery. Computational explorations of the creative processes.* Cambridge, MA: MIT Press.

Larkin, J. H., & Simon, H. A. (1987). Why a diagram is (sometimes) worth 10,000 words. *Cognitive Science, 11*, 65–100.

Lave, J. (1988). *Cognition in practice: Mind, mathematics and culture in everyday life.* Cambridge, England: Cambridge University Press.

Lave, J., & Wenger, E. (1991). *Situated learning: Legitimate peripheral participation.* Cambridge, England: Cambridge University Press.

Lawson, B. R. (1984). Cognitive strategies in architectural design. In N. Cross (Ed.), *Developments in design methodology* (pp. 209–220). Chichester, England: Wiley. (Original work published 1979)

Lawson, B. R. (1994). *Design in mind*. London: Butterworth.

Le Ny, J.-F. (1979). *La sémantique psychologique* [Semantic psychology]. Paris: Presses Universitaires de France.

Le Ny, J.-F. (1989a). Questions ouvertes sur la localisation [Open questions about localisation]. *Intellectica*, *2*(8), 61–84.

Le Ny, J.-F. (1989b). *Science cognitive et compréhension du langage* [Cognitive science and verbal understanding]. Paris: Presses Universitaires de France.

Lebahar, J.-C. (1983). *Le dessin d'architecte. Simulation graphique et réduction d'incertitude* [Architectural drawing. Graphic simulation and uncertainty reduction]. Roquevaire, France: Editions Parenthèses.

Leclercq, P. (1999). *Interpretative tool for architectural sketches*. In J. S. Gero & B. Tversky (Eds.), Visual and spatial reasoning in design. Sydney, Australia: University of Sydney, Key Centre of Design Computing and Cognition.

Lee, A., & Pennington, N. (1994). The effects of paradigms on cognitive activities in design. *International Journal of Human–Computer Studies*, *40*, 577–601.

Leplat, J. (1981). Task analysis and activity analysis in situations of field diagnosis. In J. Rasmussen & W. B. Rouse (Eds.), *Human detection and diagnostic of system failures* (pp. 289–300). New York: Plenum.

Leplat, J. (1992a). L'analyse psychologique du travail [The psychological analysis of work]. In J. Leplat (Ed.), *L'analyse du travail en psychologie ergonomique. Recueil de textes* (Vol. 1, pp. 23–39). Toulouse, France: Octarès. (Original work published 1986)

Leplat, J. (Ed.). (1992b). *L'analyse du travail en psychologie ergonomique. Recueil de textes* [Work analysis in ergonomic psychology. A collection of texts] (Vol. 1). Toulouse, France: Octarès.

Leplat, J. (Ed.). (1992c). *L'analyse du travail en psychologie ergonomique. Recueil de textes* [Work analysis in ergonomic psychology. A collection of texts] (Vol. 2). Toulouse, France: Octarès.

Letovsky, S. (1986). *Cognitive processes in program comprehension*. In E. Soloway & S. Iyengar (Eds.), *Empirical Studies of Programmers: First Workshop (ESP1)*. Norwood, NJ: Ablex.

Lieber, J., & Napoli, A. (1997, May). *Planification à partir de cas et classification* [Case-based planning and classification]. Paper presented at JICAA'97, Journées ingénierie des connaissances et apprentissage automatique, Roscoff, France.

Lindemann, U. (Ed.). (2003). *Human behaviour in design*. Berlin, Germany: Springer.

Logan, B., & Smithers, T. (1993). Creativity and design as exploration. In J. S. Gero & M. L. Maher (Eds.), *Modeling creativity and knowledge-based design* (pp. 193–175). Hillsdale, NJ: Lawrence Erlbaum Associates.

Love, T. (1999). Computerising affective design cognition. *International Journal of Design Computing*, *2*. Available at http://www.arch.usyd.edu.au/kcdc/journal/vol2/love/cadcmain.htm

Löwgren, J. (1995). Applying design methodology to software development. In G. M. Olson & S. Schuon (Eds.), *Symposium on Designing interactive systems. Proceedings of the Conference on Designing Interactive Systems: Processes, practices, methods, and techniques (DIS'95)* (pp. 87–95). New York: ACM Press.

Malhotra, A., Thomas, J. C., Carroll, J. M., & Miller, L. A. (1980). Cognitive processes in design. *International Journal of Man-Machine Studies*, *12*, 119–140.

Marmaras, N., & Pavard, B. (1999). Problem-driven approach to the design of information technology systems supporting complex cognitive tasks. *Cognition, Technology & Work*, *1*, 222–236.

Marples, D. L. (1961). The decisions of engineering design. *IRE Transactions on Engineering Management*, *June*, 55–70.

Martin, G., Détienne, F., & Lavigne, E. (2000, July). *Negotiation in collaborative assessment of design solutions: An empirical study on a concurrent engineering process*. Paper presented at the International Conference on Concurrent Engineering, CE'2000, Lyon, France.

Martin, G., Détienne, F., & Lavigne, E. (2001, July). *Analysing viewpoints in design through the argumentation process*. Paper presented at Interact 2001, Tokyo, Japan.

Mayer, R. E. (1989). Human nonadversary problem solving. In K. J. Gilhooly (Ed.), *Human and machine problem solving* (pp. 39–56). New York: Plenum.

Mayer, R. E. (1999). Fifty years of creativity research. In R. J. Sternberg (Ed.), *Handbook of creativity* (pp. 449–460). Cambridge, England: Cambridge University Press.

McGown, A., Green, G., & Rodgers, P. A. (1998). Visible ideas: Information patterns of conceptual sketch activity. *Design Studies, 19*(4), 431–453.

McGrath, J. E. (1984). *Groups: Interaction and performance.* Englewood Cliffs, NJ: Prentice-Hall.

McMahon, T. (1988). Is reflective practice synonymous with action research? *Educational Action Research, 7,* 163–168.

McNeill, T., Gero, J. S., & Warren, J. (1998). Understanding conceptual electronic design using protocol analysis. *Research in Engineering Design, 10,* 129–140.

Méhier, C. (2005). Intégrer des contraintes environnementales dans la conception. L'éco-conception [Integrate environnemental constraints in design. Eco-design]. In J. Forest, C. Méhier, & J.-P. Micaëlli (Eds.), *Pour une science de la conception. Fondements, méthodes, pratiques* (pp. 93–117). Belfort-Montbéliard, France: Université de Technologie de Belfort-Montbéliard.

Michard, A. (1982). Graphical presentation of Boolean expressions in a database query language: Design notes and an ergonomic evaluation. *Behaviour and Information Technology, 1*(3), 279–288.

Millen, D. R. (2000). Rapid ethnography: Time deepening strategies for HCI field research. In D. Boyarski & W. A. Kellogg (Eds.), *Proceedings of the Conference on Designing Interactive Systems: Processes, practices, methods, and techniques (DIS'00)* (pp. 280–286). New York: ACM Press.

Miller, G. A., Galanter, E., & Pribram, K. H. (1960). *Plans and the structure of behaviour.* New York: Holt.

Minsky, M. (1961). Steps toward artificial intelligence. *Proceedings IRE, 49,* 8–30.

Mollo, V., & Falzon, P. (2004). Auto- and allo-confrontation as tools for reflective activities. *Applied Ergonomics, 35*(6), 531–540.

Moore, S., & Oaksford, M. (Eds.). (2002). *Emotional cognition: From brain to behavior.* Philadelphia: John Benjamins.

Morais, A., & Visser, W. (1985). *Etude exploratoire de la programmation d'automates industriels chez des élèves de l'enseignement technique* [An exploratory study on the programming of industrial programmable controllers by control engineering students] (Research Rep. No. 404). Rocquencourt, France: Institut National de Recherche en Informatique et en Automatique. Available at http://www.inria.fr/rrrt/rr-0404.html

Moran, T. P., & Carroll, J. M. (Eds.). (1996a). *Design rationale: Concepts, techniques, and use.* Mahwah, NJ: Lawrence Erlbaum Associates.

Moran, T. P., & Carroll, J. M. (1996b). Overview of design rationale. In T. P. Moran & J. M. Carroll (Eds.), *Design rationale: Concepts, techniques, and use* (pp. 1–19). Mahwah, NJ: Lawrence Erlbaum Associates.

Nakakoji, K., Yamamoto, Y., Takada, S., & Reeves, B. N. (2000). Two-dimensional spatial positioning as a means for reflection in design. In D. Boyarski & W. A. Kellogg (Eds.), *Proceedings of the Conference on Designing Interactive Systems: Processes, practices, methods, and techniques (DIS'00)* (pp. 145–154). New York: ACM Press.

Nardi, B. A. (1996). Studying context: A comparison of activity theory, situated action models, and distributed cognition. In B. A. Nardi (Ed.), *Context and consciousness. Activity theory and human–computer interaction* (pp. 69–102). Cambridge, MA: MIT Press.

Navinchandra, D. (1991). *Exploration and innovation in design: Towards a computational model.* New York: Springer.

Neiman, B., Gross, M. D., & Do, E. Y.-L. (1999, August). *Sketches and their functions in early design. A retrospective analysis of a pavilion house.* Paper presented at the 4th Design Thinking Research Symposium, DTRS99, Cambridge, MA.

Newell, A. (1969). Heuristic programming: Ill-structured problems. In J. Aronofsky (Ed.), *Progress in operations research: Relationship between operations research and the computer* (Vol. 3, pp. 360–414). [Also in P. S. Rosenbloom, J. E. Laird, & A. Newell (Eds.) (1993), *The Soar papers: Research on integrated intelligence* (Vol. 1, pp. 3–54). London: MIT Press.]

Newell, A. (1990). *Unified theories of cognition.* Cambridge, MA: Harvard University Press.

Newell, A., Shaw, J. C., & Simon, H. A. (1957a, February). *Empirical explorations of the logic theory machine: A case study in heuristics.* Paper presented at the Western Joint Computer Conference, Los Angeles, CA.

Newell, A., Shaw, J. C., & Simon, H. A. (1957b). Problem solving in humans and computers. *Carnegie Technical, 21*(4), 35–38.

Newell, A., Shaw, J. C., & Simon, H. A. (1958). Elements of a theory of human problem solving. *Psychological Review, 65,* 151–166.

Newell, A., & Simon, H. A. (1956). The logic theory machine. *IRE Transactions on Information Theory, IT, 2*(3), 61–79.

Newell, A., & Simon, H. A. (1972). *Human problem solving.* Englewood Cliffs, NJ: Prentice-Hall.

Newell, A., & Simon, H. A. (1976). Computer science as empirical inquiry: Symbols and search. [1975 ACM Turing Award lecture]. *Communications of the Association for Computing Machinery, 19*(3), 113–126.

Newman, M. W., & Landay, J. A. (2000). Sitemaps, storyboards, and specifications: A sketch of web site design practice. In D. Boyarski & W. A. Kellogg (Eds.), *Proceedings of the Conference on Designing Interactive Systems: Processes, practices, methods, and techniques (DIS'00)* (pp. 263–274). New York: ACM Press.

Nii, H. P. (1986a). Blackboard systems: The blackboard model of problem solving and the evolution of blackboard architectures. Part One. *AI Magazine, 7*(2), 38–53.

Nii, H. P. (1986b). Blackboard systems: Blackboard application systems and a knowledge engineering perspective. Part Two, *AI Magazine, 7*(3), 82–107.

Nilsson, M. (2005, August). *Workplace studies revisited.* Paper presented at IRIS 28, the 28th Information systems research seminar in Scandinavia, Kristiansand, Norway. Retrieved January 11, 2006, from www.hia.no/iris28/Docs/IRIS2028-1023.pdf

Nisbett, R. E., & Wilson, T. D. (1977). Telling more than we can know: Verbal reports on mental processes. *Psychological Review, 84,* 231–259.

Norman, D. A. (1991). Cognitive artifacts. In J. M. Carroll (Ed.), *Designing interaction: Psychology of the human–computer interface* (pp. 17–38). New York: Cambridge University Press.

Norman, D. A. (1993). Cognition in the head and in the world: An introduction to the special issue on Situated action. *Cognitive Science, 17*(1), 1–6.

Norman, D. A. (1996). Design as practiced. In T. Winograd (Ed.), *Bringing design to software* (pp. 233–247). New York: ACM Press.

Norman, D. A., & Draper, S. W. (Eds.). (1986). *User centered system design. New perspectives on human–computer interaction.* Hillsdale, NJ: Lawrence Erlbaum Associates.

Ochanine, D. (1978). Le rôle des images opératives dans la régulation des activités de travail [The role of operative images in work activity regulation]. *Psychologie et Education, 2, 63-72,*

Okada, T., & Simon, H. A. (1997). Collaborative discovery in a scientific domain. *Cognitive Science, 21*(2), 109–146.

Olson, G. M., Olson, J. S., Carter, M. R., & Storrøsten, M. (1992). Small group design meetings: An analysis of collaboration. *Human–Computer Interaction, 7,* 347–374.

Olson, G. M., Olson, J. S., Storrøsten, M., Carter, M. R., Herbsleb, J. D., & Rueter, H. (1996). The structure of activity during design meetings. In T. P. Moran & J. M. Carroll (Eds.), *Design rationale: Concepts, techniques and uses* (pp. 217–239). Mahwah, NJ: Lawrence Erlbaum Associates.

Olson, G. M., Sheppard, S., & Soloway, E. (Eds.). (1987). *Empirical Studies of Programmers: Second Workshop (ESP2).* Norwood, NJ: Ablex.

Ormerod, T. C. (1990). Human cognition and programming. In J.-M. Hoc, T. R. G. Green, R. Samurçay, & D. J. Gilmore (Eds.), *Psychology of programming* (pp. 63–82). London: Academic Press.

Oxman, R. (2004). Think-maps: Teaching design thinking in design education. *Design Studies, 25*(1), 63–91.

Pahl, G., Badke-Schaub, P., & Frankenberger, E. (1999). Resume of 12 years interdisciplinary empirical studies of engineering design in Germany. *Design Studies, 20*(5), 481–494.

Pahl, G., & Beitz, W. (1977). *Konstruktionslehre* [Theory of design]. Berlin, Germany: Springer.

Pahl, G., & Beitz, W. (1984). *Engineering design* (K. M. Wallace, Trans.). London: The Design Council.

Pahl, G., & Beitz, W. (1996). *Engineering design. A systematic approach* (K. M. Wallace, L. Blessing, & F. Bauert, Trans.; 2nd, enlarged, and updated ed.). London: Springer.

Pahl, G., Frankenberger, E., & Badke-Schaub, P. (1999). Historical background and aims of interdisciplinary research between Bamberg, Darmstadt and Munich. *Design Studies, 20*(5), 401–406.

Pakman, M. (2000). Thematic foreword: Reflective practices: The legacy of Donald Schön. *Cybernetics and Human Knowing, 7*(2–3), 5–8.

Papantonopoulos, S. (2004). How system designers think: A study of design thinking in human factors engineering. *Ergonomics, 47*(14), 1528–1548.

Paton, R., & Neilson, I. (Eds.). (1999). *Visual representations and interpretations.* London: Springer.

Pennington, N., & Grabowski, B. (1990). The tasks of programming. In J.-M. Hoc, T. R. G. Green, R. Samurçay, & D. J. Gilmore (Eds.), *Psychology of programming* (pp. 145–162). London: Academic Press.

Perrin, J. (1999). Diversité des représentations du processus de conception, diversité des modes de pilotage de ces processus [Diversity of design-process representations, diversity of process-piloting modes]. In J. Perrin (Ed.), *Pilotage et évaluation des processus de conception* (pp. 19–39). Paris: L'Harmattan.

Perrin, J. (Ed.). (2002). *International Conference in honour of Herbert Simon, "The Sciences of Design. The Scientific Challenge for the 21st Century"* (INSA, Lyon, France, March 15–16, 2002).

Peterson, D. (Ed.). (1996). *Forms of representation: An interdisciplinary theme for cognitive science.* Exeter, England: Intellect Books.

Pitrat, J. (2002a). Herbert Simon, pionnier de l'intelligence artificielle [Herbert Simon, pioneer of artificial intelligence]. *Revue d'Intelligence Artificielle, 16*(1–2), 11–16.

Pitrat, J. (Ed.). (2002b). Représentations, découverte et rationalité. Hommage à Herbert Simon [Numéro spécial][Representations, discovery and rationality. Hommage to Herbert Simon, Special issue]. *Revue d'Intelligence Artificielle, 16*(1–2).

Pu, P. (1993). Introduction: Issues in case-based design systems. *AI EDAM, 7*(2), 79–85.

Purcell, T. (1998a). Editorial [to the Special issue on Sketching and drawing in design]. *Design Studies, 19*(4), 385–387.

Purcell, T. (Ed.). (1998b). Sketching and drawing in design [Special issue]. *Design Studies, 19*(4).

Purcell, T., & Gero, J. S. (1996). Design and other types of fixation. *Design Studies, 17*(4), 363–383.

Qin, Y., & Simon, H. A. (1990). Laboratory replication of scientific discovery processes. *Cognitive Science, 14*, 281–312.

Rasmussen, J. (1986). *Information processing and human-machine interaction: An approach to cognitive engineering.* Amsterdam: North-Holland.

Ratcliff, B., & Siddiqi, J. I. A. (1985). An empirical investigation into problem decomposition strategies used in program design. *International Journal of Man-Machine Studies, 22*, 77–90.

Reed, S. (1993). Imagery and discovery. In B. Roskos-Ewoldsen, M. Intons-Peterson, & R. Anderson (Eds.), *Imagery, creativity and discovery: A cognitive perspective* (pp. 287–312). Amsterdam: North-Holland.

Reimann, P., & Chi, M. T. H. (1989). Human expertise. In K. J. Gilhooly (Ed.), *Human and machine problem solving* (pp. 161–191). New York: Plenum.

Reitman, W. (1964). Heuristic decision procedures, open constraints, and the structure of ill-defined problems. In M. W. Shelley & G. L. Bryan (Eds.), *Human judgments and optimality* (pp. 282–315). New York: Wiley.

Reitman, W. (1965). *Cognition and thought.* New York: Wiley.

Relieu, M., Salembier, P., & Theureau, J. (2004). Introduction au numéro spécial "Activité et action/Cognition Située" [Introduction to the special issue "Activity and action/Situated cognition"]. @ctivités. revue électronique, 1(2), 3–10. Available at http://www.activites.org/v1ln12/intro.pdf

Representation [Special issue]. (1995). Cognitive Systems, 4(2).

Ribert-Van De Weerdt, C. (2003). Intérêts et difficultés de l'analyse des émotions en psycho-ergonomie [Interests and difficulties of emotions analysis in psycho-ergonomics]. Psychologie Française, No. spécial "Recherches en psychologie ergonomique," 48(2), 9–16.

Richard, J.-F. (1990). Les activités mentales. Comprendre, raisonner, trouver des solutions [Mental activities. Understanding, reasoning, finding solutions]. Paris: Armand Colin.

Richard, J.-F., Poitrenaud, S., & Tijus, C. (1993). Problem-solving restructuration: Elimination of implicit constraints. Cognitive Science, 17, 497–529.

Rist, R. S. (1990). Variability in program design: The interaction of process with knowledge. International Journal of Man-Machine Studies, 33(3), 305–322.

Rist, R. S. (1991a). Knowledge creation and retrieval in program design: A comparison of novice and intermediate student programmers. Human–Computer Interactions, 6, 1–46.

Rist, R. S. (1991b, August). Models of routine and non-routine design in the domaine of programming. In J. S. Gero & F. Sudweeks (Eds.), Preprints of the "Artificial Intelligence in Design" Workshop of the Twelfth International Joint Conference on Artificial Intelligence, Sydney, Australia, 25 August. Sydney, Australia: University of Sydney.

Rittel, H. W. J. (1984). Second-generation design methods (Interview with Donald P. Grant and Jean-Pierre Protzen). In N. Cross (Ed.), Developments in design methodology (pp. 317–327). Chichester, England: Wiley. (Original work published 1972)

Rittel, H. W. J., & Webber, M. M. (1984). Planning problems are wicked problems. In N. Cross (Ed.), Developments in design methodology (pp. 135–144). Chichester, England: Wiley. (Original work published 1973)

Robert, J.-M. (1979). Les résultats des recherches sur le processus de conception, les comportements et les processus cognitifs mis en jeu [Research results concerning the design process, the behavior and cognitive processes implemented] (INRIA Tech. Rep. No. EC 7912 R01). Rocquencourt, France: Institut National de Recherche en Informatique et en Automatique.

Robinson, M., & Bannon, L. J. (1991, September). Questioning representations. Paper presented at ECSCW'91, Amsterdam.

Rodgers, P. A., Green, G., & McGown, A. (2000). Using concept sketches to track design progress. Design Studies, 21(5), 451–464.

Roozenburg, N. F. M., & Dorst, K. (1999). Describing design as a reflective practice: Observations on Schön's theory of practice. In E. Frankenberger, P. Badke-Schaub, & H. Birkhofer (Eds.), Designers. The key to successful product development (pp. 29–41). London: Springer.

Rosch, E. (1978). Principles of categorization. In E. Rosch & B. B. Lloyd (Eds.), Cognition and categorization (pp. 27–48). Hillsdale, NJ: Lawrence Erlbaum Associates.

Rosson, M. B., & Alpert, S. R. (1990). The cognitive consequences of object-oriented design. Human–Computer Interaction, 5, 345–379.

Rowe, P. G. (1987). Design thinking. Cambridge, MA: MIT Press.

Rumelhart, D. E. (1989). Toward a microstructural account of human reasoning. In S. Vosniadou & A. Ortony (Eds.), Similarity and analogical reasoning (pp. 298–312). Cambridge, England: Cambridge University Press.

Ryle, G. (1973). The concept of mind. Harmondsworth, England: Penguin (Original work published 1949).

Saunders, R. (2001). Design Thinking by Peter G. Rowe [Book review]. Retrieved September 7, 2005, from http://www.arch.usyd.edu.au/~rob/study/DesignThinking.html

Scacchi, W. (2001). Process models in software engineering. In J. J. Marciniak (Ed.), Encyclopedia of software engineering (2nd ed., pp. 993–1005). New York: Wiley.

Scaife, M., & Rogers, Y. (1996). External cognition: How do graphical representations work? *International Journal of Human-Computer Studies*, *45*(1), 185-213.

Scharmer, C. O. (1999). *Primary knowing: When perception happens from the whole field. Interview with Professor Eleanor Rosch, Berkeley, CA, October 15, 1999.* Retrieved July 12, 2004, from http://www.dialogonleadership.org/interviewRosch.html

Schmidt, K., & Wagner, I. (2002). Coordinative artifacts in architectural practice. In M. Blay-Fornarino, A. M. Pinna-Dery, K. Schmidt, & P. Zaraté (Eds.), Cooperative systems design: A challenge of the mobility age (pp. 257–274). Amsterdam: IOS Press.

Schön, D. A. (1983). *The reflective practitioner: How professionals think in action.* New York: Basic Books. (Reprinted 1995)

Schön, D. A. (1984). *The design studio: An exploration of its traditions and potentials.* London: RIBA.

Schön, D. A. (1987a, April). *Educating the reflective practitioner.* Paper presented at the Annual Meeting of the American Educational Research Association, Washington, DC. Retrieved March 15, 2006, from http://educ.queensu.ca/~russellt/forum/schon87.htm

Schön, D. A. (1987b). *Educating the reflective practitioner.* San Francisco: Jossey-Bass.

Schön, D. A. (1988). Designing: Rules, types and worlds. *Design Studies*, *9*(3), 181–190.

Schön, D. A. (1992). Designing as reflective conversation with the materials of a design situation. *Knowledge-Based Systems*, *5*(1), 3–14.

Schön, D. A., & Wiggins, G. (1992). Kinds of seeing and their functions in designing. *Design Studies*, *13*(2), 135–156.

Science of Design. Program Solicitation NSF 04–552. National Science Foundation (*2004.* last updated September 6, 2005). Retrieved October 12, 2005, from http://www.nsf.gov/pubs/2004/nsf04552/nsf04552.htm

Scrivener, S. A. R. (1997). Drawing from visual thinking. *TRACEY electronic journal, Issue "Thinking."* Retrieved April 5, 2005, from http://www.lboro.ac.uk/departments/ac/tracey/thin/scriven.html

Seaman, C. B., & Basili, V. R. (1998). Communication and organization: An empirical study of discussion in inspection meetings. *IEEE Transactions on Software Engineering*, *24*(6), 559–572.

Sharpe, J. E. E. (1995). Computer tools for integrated conceptual design. *Design Studies*, *16*(4), 471–488.

Sim, S. K., & Duffy, A. H. B. (2003). Towards an ontology of generic engineering design activities. *Research in Engineering Design*, *14*(4), 200–223.

Simon, H. A. (1955). A behavioral model of rational choice. *Quarterly Journal of Economics*, *69*, 99–118.

Simon, H. A. (1973). The structure of ill-structured problems. *Artificial Intelligence*, *4*, 181–201. [Also in N. Cross (Ed.) (1984), *Developments in design methodology* (pp. 145–166). Chichester, England: Wiley.]

Simon, H. A. (1975). Style in design. In C. Eastman (Ed.), *Spatial synthesis in computer-aided building design* (pp. 287–309). London: Applied Science Publishers. [The version of Simon's text we refer to was first published in J. Archea & C. Eastman (Eds.) (1971). *EDRA TWO, Proceedings 2nd Ann. Environmental Design Research Association Conference, October 1–10, 1970* (pp. 1–10). Stroudsbury, PA: Dowden, Hutchinson, & Ross, Inc.]

Simon, H. A. (1977a). *Models of discovery.* Boston: Reidel.

Simon, H. A. (1977b). The next hundred years: Engineering design. In L. E. Jones (Ed.), *The next hundred years* (pp. 89–104). Toronto: University of Toronto, Faculty of Applied Science and Engineering.

Simon, H. A. (1978). Information-processing theory of human problem solving. In W. K. Estes (Ed.), *Handbook of learning and cognitive processes* (Vol. 5, pp. 271–295). Hillsdale, NJ: Lawrence Erlbaum Associates.

Simon, H. A. (1979). Information processing models of cognition. *Annual Review of Psychology*, *30*, 363–396.

Simon, H. A. (1980). Technology: Source of opportunity and constraint in design. *College of Design, Architecture, and Art Journal, 1*, 26–33.

Simon, H. A. (1982). *Models of bounded rationality* (Vols. 1 & 2). Cambridge, MA: MIT Press.

Simon, H. A. (1987). Making management decisions: The role of intuition and emotion. *Academy of Management EXECUTIVE, 1*(1), 57–64.

Simon, H. A. (1992a). Alternative representations for cognition: Search and reasoning. In H. L. Pick, Jr., P. van den Broek, & D. C. Knill (Eds.), *Cognition: Conceptual and methodological issues* (pp. 121–142). Washington, DC: American Psychological Association.

Simon, H. A. (1992b). Rational decision-making in business organizations [Nobel Memorial Lecture, December 8, 1978]. In A. Lindbeck (Ed.), *Nobel lectures, economics 1969–1980* (pp. 343–371). Singapore: World Scientific.

Simon, H. A. (1992c). Scientific discovery as problem solving. *International Studies in the Philosophy of Science, 6*, 3–14.

Simon, H. A. (1992d). Scientific discovery as problem solving: Reply to critics. *International Studies in the Philosophy of Science, 6*, 69–88.

Simon, H. A. (1995a, August). *Explaining the ineffable: AI on the topics of intuition, insight and inspiration.* Paper presented at the Fourteenth International Joint Conference on Artificial Intelligence, Montreal, Quebec, Canada.

Simon, H. A. (1995b). Problem forming, problem finding, and problem solving in design. In A. Collen & W. W. Gasparski (Eds.), *Design and systems: General applications of methodology* (Vol. 3, pp. 245–257). New Brunswick, NJ: Transaction. (Text of a lecture delivered to the First International Congress on Planning and Design Theory, Boston, 1987)

Simon, H. A. (1997a). Integrated design and process technology. *Journal of Integrated Design and Process Science, 1*(1), 9–16.

Simon, H. A. (1997b). *Models of bounded rationality* (Vol. 3). Cambridge, MA: MIT Press.

Simon, H. A. (1999a). Foreword. In C. L. Dym & P. Little (1999). *Engineering design* (pp. vii–viii). New York: Wiley.

Simon, H. A. (1999b). *The sciences of the artificial* (3rd, rev. ed.). Cambridge, MA: MIT Press. (Original work published 1969)

Simon, H. A. (2000). Bounded rationality in social science: Today and tomorrow. *Mind & Society, 1*(1), 25–39.

Simon, H. A. (2001a). Creativity in the arts and the sciences. *The Kenyon Review, Spring, 23*(2), 203–220.

Simon, H. A. (2001b). "Seek and ye shall find": How curiosity engenders discovery. In K. D. Crowley, C. D. Schunn, & T. Okada (Eds.), *Designing for science: Implications from everyday classroom, and professional settings* (pp. 3–18). Mahwah, NJ: Lawrence Erlbaum Associates.

Simon, H. A. (2002). Sur le colloque Sciences de l'Intelligence, Sciences de l'Artificiel. Extraits des commentaires et des réponses aux questions [About the Conference Sciences of Intelligence, Sciences of the Artificial. Excerpts of the comments et answers to the questions]. *Revue d'Intelligence Artificielle, 16*(1–2), 39–52. (The conference took place in 1984)

Simon, H. A., & Newell, A. (1956). Models: Their uses and limitations. In L. D. White (Ed.), *The state of the social sciences* (pp. 66–83). Chicago: University of Chicago Press.

Situated Action [Special issue]. (1993). *Cognitive Science, 17*(1).

Smith, G. J., & Gero, J. S. (2005). What does an artificial design agent mean by being 'situated'? *Design Studies, 26*(5), 535–561.

Smith, M. K. (2001, July). *Donald schon (schön): Learning, reflection and change.* Retrieved August 23, 2001, from http://www.infed.org/thinkers/et-schon.htm

Smyth, B., & Keane, M. T. (1995). Some experiments on adaptation-guided retrieval. In M. Veloso & A. Aamodt (Eds.), *Case-based reasoning, Research and development, First International Conference, Proceedings of ICCBR-95* (pp. 313–324). New York: Springer.

Soloway, E., & Iyengar, S. (Eds.). (1986). *Empirical Studies of Programmers: First Workshop (ESP1).* Norwood, NJ: Ablex.

Sosa, R., & Gero, J. S. (2003, August). *Design and change: A model of situated creativity*. Paper presented at the "Creative systems. Approaches to creativity in artificial intelligence and cognitive science" Workshop held in conjunction with IJCAI'03, the Eighteenth International Joint Conference on Artificial Intelligence, Acapulco, Mexico.

Stacey, M., & Eckert, C. (2003). Against ambiguity. *Computer Supported Cooperative Work, 12*(2), 153–183.

Stahl, G. (2002, November). *Contributions to a theoretical framework for CSCL*. Paper presented at the CSCL'02, Computer Supported Collaborative Learning, Boulder, CO. Retrieved March 28, 2006, from http://newmedia.colorado.edu/cscl/81.pdf

The standard waterfall model for systems development. (n.d.; last updated January 29, 2004). Retrieved October 12, 2005, from http://asd-www.larc.nasa.gov/barkstrom/public/The_ Standard_Waterfall_Model_For_Systems_Development.htm

Star, S. L. (1988). The structure of ill-structured solutions: Heterogeneous problem-solving, boundary objects and distributed artificial intelligence. In M. Huhns & L. Gasser (Eds.), *Distributed artificial intelligence* (Vol. 3, pp. 37–54). Los Altos, CA: Morgan Kaufman.

Staufer, L. A., & Ullman, D. G. (1988). A comparison of the results of empirical studies into the mechanical design process. *Design Studies, 9*(2), 107–114.

Stefik, M. J. (1981a). Planning and meta-planning (MOLGEN: Part 2). *Artificial Intelligence, 16*, 141–170.

Stefik, M. J. (1981b). Planning with constraints (MOLGEN: Part 1). *Artificial Intelligence, 16*, 111–140.

Steinberg, L. I. (1987, July). *Design as refinement plus constraint propagation: The VEXED experience*. Paper presented at the Sixth National Conference on Artificial Intelligence (AAAI-87), Seattle, WA.

Stempfle, J., & Badke-Schaub, P. (2002). Thinking in design teams. An analysis of team communication. *Design Studies, 23*(5), 473–496.

Sternberg, R. J. (Ed.). (1999). *Handbook of creativity*. Cambridge, England: Cambridge University Press.

Stolterman, E. (1991). How system designers think about design and methods. Some reflections based on an interview study. *Scandinavian Journal of Information Systems, 3*, 137–150. (Retrieved October, 116, 2002, from http://iris.informatik.gu.se/sjis/vol2004/stolter.shtml)

Suchman, L. A. (1987). *Plans and situated actions. The problem of human-machine communication*. Cambridge, England: Cambridge University Press.

Suchman, L. A. (1993). Response to Vera and Simon's situated action: A symbolic interpretation. *Cognitive Science, 17*(1), 71–77.

Suchman, L. A. (1995). Making work visible. *Communications of the ACM, 38*(9), 56–64.

Suwa, M., & Tversky, B. (1997). What do architects and students perceive in their design sketches? A protocol analysis. *Design Studies, 18*(4), 385–404.

Swaine, M. (1996). *Design: Whose job is it, anyway? Dr. Dobb's Journal of Software Tools (DDJ)*. Retrieved October 12, 2005, from http://www.ddj.com/documents/s=957/ddj9609k/9609k.htm

Tang, J. C. (1991). Findings from observational studies of collaborative work. *International Journal of Man-Machine Studies, 34*, 143–160.

Tang, J. C., & Leifer, L. J. (1988, September). *A framework for understanding the workspace activity of design teams*. Paper presented at CSCW 88, 2nd ACM Conference on Computer Supported Cooperative Work, Portland, OR.

Thomas, J. C. (1989). Problem solving by human-machine interaction. In K. J. Gilhooly (Ed.), *Human and machine problem solving* (pp. 317–362). New York: Plenum.

Thomas, J. C., & Carroll, J. M. (1979). The psychological study of design. *Design Studies, 1*(1), 5–11. [Also in N. Cross (Ed.) (1984), *Developments in design methodology* (pp. 221–235). Chichester, England: Wiley.]

Thunem, S., & Sindre, G. (1992, October). *Development with and for reuse*. Paper presented at the ERCIM 92 "Methods and Tools for Software Reuse," Heraklion, Greece.

Traverso, V., & Visser, W. (2003). Confrontation de deux méthodologies d'analyse de situations d'élaboration collective de solution [Confrontation of two methodologies for the analysis of collective solution-elaboration situations]. In J. M. C. Bastien (Ed.), *Deuxièmes Journées d'Etude en Psychologie ergonomique-EPIQUE 2003* (pp. 241–246). Rocquencourt, France: Institut National de Recherche en Informatique et en Automatique.

Tseng, W., Scrivener, S. A. R., & Ball, L. J. (2002). The impact of functional knowledge on sketching. In T. Kavanagh & T. Hewett (Eds.), *Creativity and Cognition 2002* (pp. 57–64). New York: ACM Press.

Tulving, E. (1972). Episodic and semantic memory. In E. Tulving & W. Donaldson (Eds.), *Organization of memory* (pp. 381–403). New York: Academic Press.

Tulving, E. (1983). *Elements of episodic memory.* Oxford, England: Oxford University Press.

Tversky, A., & Kahneman, D. (1981). The framing of decisions and the psychology of choice. *Science, 211,* 453–458.

Tweed, C. (1999). Learning to see architecturally. In R. Paton & I. Neilson (Eds.), *Visual representations and interpretations* (pp. 232–246). London: Springer.

Tzonis, A. (1992). Huts, ships and bottleracks: Design by analogy for architects and/or machines. In N. Cross, K. Dorst, & N. F. M. Roozenburg (Eds.), *Research in design thinking* (pp. 139–164). Delft, Netherlands: Delft University Press.

Ullman, D. G. (1992). *The mechanical design process.* New York: McGraw-Hill.

Ullman, D. G., & Culley, S. J. (1994). The mechanical design process. *Design Studies, 15*(1), 115.

Ullman, D. G., Dietterich, T. G., & Staufer, L. A. (1988). A model of the mechanical design process based on empirical data. *AI EDAM, 2,* 33–52.

Ullman, D. G., Staufer, L. A., & Dietterich, T. G. (1987). Toward expert CAD. *Computers in Mechanical Engineering, 6*(3), 56–70.

Ullman, D. G., Wood, S., & Craig, D. L. (1990). The importance of drawing in the mechanical design process. *Computing and Graphics, 14*(2), 263–274.

Valkenburg, R. (2001). Schön revised: Describing team designing with reflection-in-action. In P. Lloyd & H. Christiaans (Eds.), *Proceedings of DTRS 5* (pp. 315–329). Delft, Netherlands: Delft University Press. [Consulted in its Internet version, no longer accessible.]

Valkenburg, R., & Dorst, K. (1998). The reflective practice of design teams. *Design Studies, 19*(3), 249–271.

Van der Lugt, R. (2000). Developing a graphic tool for creative problem solving in design groups. *Design Studies, 21*(5), 505–522.

Van der Lugt, R. (2002). Functions of sketching in design idea generation meetings. In T. Kavanagh & T. Hewett (Eds.), *Creativity and Cognition 2002* (pp. 72–79). New York: ACM Press.

Van der Lugt, R. (2003, August). *Relating the quality of the idea generation process to the quality of the resulting design ideas.* Paper presented at the International Conference on Engineering Design, ICED'03, Stockholm, Sweden.

Van Someren, M. W., Reimann, P., Boshuizen, H. P. A., & De Jong, T. (1988). *Learning with multiple representations.* Amsterdam: Elsevier.

VDI (1987). *Systematic approach to the design of technical systems and products. Guideline VDI 2221* (K. M. Wallace, Trans.). Berlin, Germany: Beuth.

Vera, A. H., & Simon, H. A. (1993a). Situated action: Reply to reviewers. *Cognitive Science, 17*(1), 77–86.

Vera, A. H., & Simon, H. A. (1993b). Situated action: A symbolic interpretation. *Cognitive Science, 17*(1), 7–48.

Vera, A. H., & Simon, H. A. (1993c). Situated action: Reply to William Clancey. *Cognitive Science, 17*(1), 117–133.

Vermersch, P. (1994). *L'entretien d'explicitation* [The explicitation interview]. Paris: ESF.

Verstijnen, I. M., Heylighen, A., Wagemans, J., & Neuckermans, H. (2001, July). *Sketching, analogies, and creativity. On the shared research interests of psychologists and designers.* Paper presented at the 2nd International Conference on Visual and Spatial Reasoning in Design, VR'01, Bellagio, Lake Como, Italy.

Verstijnen, I. M., Van Leeuwen, C., Goldschmidt, G., Hamel, R., & Hennessey, J. M. (1998). Sketching and creative discovery. *Design Studies, 19*(4), 519–546.

Visser, W. (1985, June). *Modélisation de l'activité de programmation de systèmes de commande* [Modeling the activity of designing control-system software]. Paper presented at the Colloque COGNITIVA 85 (Tome 2), Paris.

Visser, W. (1986a). *Activité de conception d'une installation automatisée. 1. La construction du schéma de fonctionnement. 1a. Le schéma de séquences* [Designing an automated installation. 1. Constructing the schema of functioning. 1a. The sequences schema] (Tech. Rep. of the Projet de Psychologie Ergonomique pour l'Informatique). Rocquencourt, France: Institut National de Recherche en Informatique et en Automatique.

Visser, W. (1986b). *Etude de deux étapes de la conception. Ecriture et mise au point d'un programme [Studying two stages in design. Coding, and testing and debugging a software program]* (Tech. Rep. of the Projet de Psychologie Ergonomique pour l'Informatique). Rocquencourt, France: Institut National de Recherche en Informatique et en Automatique.

Visser, W. (1986c). *Etude préliminaire du travail du programmeur d'automate programmable dans un bureau d'études électrique* [Preliminary study on the work of the programmer of an programmable controller] (Tech. Rep. of the Projet de Psychologie Ergonomique pour l'Informatique). Rocquencourt, France: Institut National de Recherche en Informatique et en Automatique.

Visser, W. (1987a, May). *Abandon d'un plan hiérarchique dans une activité de conception* [Giving up a hierarchical plan in a design activity]. Paper presented at the Colloque scientifique MARI 87 Machines et Réseaux Intelligents-COGNITIVA 87 (Tome 1), La Villette, Paris. English version available at http://www.inria.fr/rrrt/rr-0814.htm

Visser, W. (1987b). Strategies in programming programmable controllers: A field study on a professional programmer. In G. M. Olson, S. Sheppard, & E. Soloway (Eds.), *Empirical Studies of Programmers: Second Workshop (ESP2)* (pp. 217–230). Norwood, NJ: Ablex.

Visser, W. (1988a). L'activité de comparaison de représentations dans la mise au point de programmes [Comparing representations in program testing and debugging]. *Le Travail Humain, Numéro Spécial "Psychologie ergonomique de la programmation informatique," 51*(4), 351–362.

Visser, W. (1988b). *Langages de programmation dédiés. Quelques exemples dans le domaine des automates programmables industriels* [Customized programming languages. Some examples in the domain of industrial programmable controllers] (Internal report). Rocquencourt, France: Institut National de Recherche en Informatique et en Automatique.

Visser, W. (1988c, June). *Towards modelling the activity of design: An observational study on a specification stage.* Paper presented at the IFAC/IFIP/IEA/IFORS Conference on Man-Machine Systems. Analysis, Design and Evaluation, Oulu, Finland.

Visser, W. (1989). *The opportunistic use of a plan in a design activity: An empirical study of specification* (Research Rep. No. 1035). Rocquencourt, France: Institut National de Recherche en Informatique et en Automatique. Available at http://www.inria.fr/rrrt/rr-1035.html

Visser, W. (1990). More or less following a plan during design: Opportunistic deviations in specification. *International Journal of Man-Machine Studies, 33,* 247–278.

Visser, W. (1991a). The cognitive psychology viewpoint on design: Examples from empirical studies. In J. S. Gero (Ed.), *Artificial Intelligence in Design'91* (pp. 505–524). Oxford, England: Butterworth-Heinemann.

Visser, W. (1991b). Evocation and elaboration of solutions: Different types of problem-solving actions. An empirical study on the design of an aerospace artifact. In T. Kohonen & F. Fogelman-Soulié (Eds.), *Cognitiva 90. At the crossroads of artificial intelligence, cognitive science and neuroscience. Proceedings of the Third COGNITIVA Symposium* (pp. 689–696). Amsterdam: Elsevier. Available at http://hal.inria.fr/inria-00000165

Visser, W. (1991c). Planning in routine design. Some counterintuitive data from empirical studies. In J. S. Gero & F. Sudweeks (Eds.), *Preprints of the "Artificial Intelligence in Design" Workshop of the Twelfth International Joint Conference on Artificial Intelligence, Sydney, Australia, 25 August.* Sydney, Australia: University of Sydney.

Visser, W. (1992a). *Design organization: There is more to expert knowledge than is dreamed of in the planner's philosophy* (Research Rep. No. 1765). Rocquencourt, France: Institut National de Recherche en Informatique et en Automatique. Available at http://www.inria.fr/rrrt/rr-1765.html. [A slightly modified version of this text was published in J. Perrin & D. Vinck (Eds.), *The role of design in the shaping of technology. Proceedings from the COST A3 and COST A4 Workshop* (pp. 213–269). Luxembourg: Office for Official Publications of the European Communities.]

Visser, W. (1992b). Designers' activities examined at three levels: Organization, strategies & problem-solving. *Knowledge-Based Systems, 5*(1), 92–104.

Visser, W. (1992c). Use of analogical relationships between design problem-solution representations: Exploitation at the action-execution and action-management levels of the activity. *Studia Psychologica, 34*(4–5), 351–357.

Visser, W. (1992d). Use of analogical relationships between design problem-solution representations: Exploitation at the action-execution and action-management levels of the activity (Abstract). *International Journal of Psychology, 27*(3–4), 156.

Visser, W. (1993a). Collective design: A cognitive analysis of cooperation in practice. In N. F. M. Roozenburg (Ed.), *Proceedings of ICED 93, 9th International Conference on Engineering Design* (Vol. 1, pp. 385–392). Zürich, Switzerland: HEURISTA.

Visser, W. (1993b). Design & knowledge modeling-Knowledge modeling as design. *Communication and Cognition-Artificial Intelligence, 10*(3), 219–233.

Visser, W. (Ed.). (1993c). *Proceedings of the "Reuse of designs: An interdisciplinary cognitive approach" Workshop held in conjunction with IJCAI'93, the Thirteenth International Joint Conference on Artificial Intelligence, Chambéry, France, August 29, 1993.* Rocquencourt, France: Institut National de Recherche en Informatique et en Automatique.

Visser, W. (1994a). Organisation of design activities: Opportunistic, with hierarchical episodes. *Interacting with Computers, 6*(3), 239–274 (Executive summary: 235–238).

Visser, W. (1994b). Planning and organization in expert design activities. In D. J. Gilmore, R. Winder, & F. Détienne (Eds.), *User-centred requirements for software engineering environments* (pp. 25–40). Berlin, Germany: Springer.

Visser, W. (1994c, September). *Use of episodic knowledge in design problem solving.* Invited paper presented at the Delft Protocols Workshop "Analysing Design Activity," Delft, Netherlands.

Visser, W. (1994d). Workshop report of the IJCAI'93: Thirteenth International Joint Conference on Artificial Intelligence Workshop on "Reuse of designs: An interdisciplinary cognitive approach." *AI Communications, 7*(1), 64–65.

Visser, W. (1995a). Reuse of knowledge: Empirical studies. In M. Veloso & A. Aamodt (Eds.), *Case-based reasoning research and development* (pp. 335–346). Berlin, Germany: Springer.

Visser, W. (1995b). Use of episodic knowledge and information in design problem solving. *Design Studies, 16*(2), 171–187. [Also in N. Cross, H. Christiaans, & K. Dorst (Eds.) (1996), *Analysing design activity* (pp. 271–289). Chichester, Great Britain: Wiley.]

Visser, W. (1996). Two functions of analogical reasoning in design: A cognitive-psychology approach. *Design Studies, 17*, 417–434.

Visser, W. (1999). Etudes en ergonomie cognitive sur la réutilisation en conception. Quelles leçons pour le raisonnement à partir de cas ? [Cognitive-ergonomics studies on reuse in design. Which lessons can we draw for case-based reasoning?]. *Revue d'Intelligence Artificielle, 13*, 129–154.

Visser, W. (2002a). Conception individuelle et collective. Approche de l'ergonomie cognitive [Individual and collective design. The cognitive-ergonomics approach]. In M. Borillo & J.-P. Goulette (Eds.), *Cognition et création. Explorations cognitives des processus de conception* (pp. 311–327). Brussels, Belgium: Mardaga.

Visser, W. (2002b). Plans. Etude de la planification de parcours en ville [Plans. A study on urban route planning] (Project No. COG 54 du Programme Cognitique, thème "Cognition spatiale")

(Final Research Rep.). In *Cognitique. Action concertée incitative 1999–2002. Résumés des projets de recherche soutenus* (pp. 87–89). Paris: Cognitique.

Visser, W. (2002c, March). *A tribute to Simon, and some-too late-questions, by a cognitive ergonomist.* Invited talk at the International Conference in honour of Herbert Simon, "The Sciences of Design. The Scientific Challenge for the 21st Century," INSA, Lyon, France. (A modified version has been published as Research Rep. No. 5144. Rocquencourt, France: Institut National de Recherche en Informatique et en Automatique. May 2004. Available at http://www.inria.fr/rrrt/rr-5144.html)

Visser, W. (2003). Dynamic aspects of individual design activities. A cognitive ergonomics viewpoint. In U. Lindemann (Ed.), *Human behaviour in design* (pp. 87–96). Berlin, Germany: Springer.

Visser, W. (2004). *Dynamic aspects of design cognition: Elements for a cognitive model of design* (Research Rep. No. 5144). Rocquencourt, France: Institut National de Recherche en Informatique et en Automatique. Available at http://www.inria.fr/rrrt/rr-5144.html

Visser, W. (2005). Designing as construction of representations: A dynamic viewpoint in cognitive design research. *Human–Computer Interaction, 21*, 103–152.

Visser, W. (in press). Co-élaboration de solutions en conception architecturale et rôle du graphico-gestuel. Point de vue de l'ergonomie cognitive [Coelaborating architectural design solutions and the role of the graphico-gestural. The cognitive-ergonomics viewpoint]. In F. Détienne & V. Traverso (Eds.), *Méthodologies d'analyse de situations coopératives de conception. Le corpus MOSAIC* [Methodologies for analysing cooperative design situations. The MOSAIC corpus]. Nancy, France: Presses Universitaires de Nancy.

Visser, W., & Bonnardel, N. (1989). *La résolution de problèmes lors de la conception d'une antenne. Analyse de l'activité* [Solving problems in designing an antenna. An analysis of the activity] (Tech. Rep. of the Projet de Psychologie Ergonomique pour l'Informatique). Rocquencourt, France: Institut National de Recherche en Informatique et en Automatique.

Visser, W., Darses, F., & Détienne, F. (2004). Approches théoriques pour une ergonomie cognitive de la conception [Theoretical approaches for the cognitive ergonomics of design]. In J.-M. Hoc & F. Darses (Eds.), *Psychologie ergonomique. Tendances actuelles* (pp. 97–118). Paris: Presses Universitaires de France.

Visser, W., & Détienne, F. (2005, December). *Articulation entre composantes verbale et graphico-gestuelle de l'interaction dans des réunions de conception architecturale* [Articulating the verbal and graphico-gestural components of the interaction in architectural design meetings]. Paper presented at the Séminaire de conception architecturale numérique, SCAN'05. "Le rôle de l'esquisse architecturale dans le monde numérique," Charenton-le-Pont, France.

Visser, W., & Falzon, P. (1988). *Eliciting expert knowledge in a design activity: Some methodological issues* (Research Rep. No. 906). Rocquencourt, France: Institut National de Recherche en Informatique et en Automatique. Available at http://www.inria.fr/rrrt/rr-0906.html

Visser, W., & Falzon, P. (1989, September). *Eliciting the knowledge of a design expert.* Abridged proceedings. Poster sessions of HCI International'89—Third International Conference on Human–Computer Interaction, Boston.

Visser, W., & Falzon, P. (1992/1993). Catégorisation et types d'expertise. Une étude empirique dans le domaine de la conception industrielle [Categorization and types of expertise. An empirical study in the industrial-design domain]. *Intellectica, 15*, 27–53.

Visser, W., & Hoc, J.-M. (1990). Expert software design strategies. In J.-M. Hoc, T. R. G. Green, R. Samurçay, & D. J. Gilmore (Eds.), *Psychology of programming* (pp. 235–250). London: Academic Press.

Visser, W., & Morais, A. (1991). Concurrent use of different expertise elicitation methods applied to the study of the programming activity. In M. J. Tauber & D. Ackermann (Eds.), *Mental models and human–computer interaction* (Vol. 2, pp. 59–79). Amsterdam: Elsevier.

Visser, W., & Richard, J.-F. (1984). *Etude préliminaire de l'activité de programmation d'automates programmables* [Preliminary study of the programmable-controllers design activity] (Tech. Rep.

of the Projet de Psychologie Ergonomique pour l'Informatique). Rocquencourt, France: Institut National de Recherche en Informatique et en Automatique.

Visser, W., & Trousse, B. (1993). Reuse of designs: Desperately seeking an interdisciplinary cognitive approach. In W. Visser (Ed.), *Proceedings of the IJCAI Thirteenth International Joint Conference on Artificial Intelligence Workshop "Reuse of designs: An interdisciplinary cognitive approach," Chambéry, France, August 29* (pp. 1–14). Rocquencourt, France: Institut National de Recherche en Informatique et en Automatique.

Visser, W., & Wolff, M. (2003a). A cognitive approach to spatial discourse production. Combining manual and automatic analyses of route descriptions. In F. Schmalhofer & R. Young (Eds.), *Proceedings of the European Cognitive Science Society Conference (EuroCogSci03) (Osnabrück, Germany, September 10–13)* (pp. 355–360). Hillsdale, NJ: Lawrence Erlbaum Associates.

Visser, W., & Wolff, M. (2003b, June). *Route plan descriptions. A cognitivo-discursive analysis.* Poster presented at the Thirteenth Annual Meeting of the Society for Text and Discourse, Madrid, Spain.

Von Glasersfeld, E. (1981). *An introduction to radical constructivism.* Retrieved September 21, 2005, from http://srri.nsm.umass.edu/vonGlasersfeld/onlinePapers/html/082.html

Von Glasersfeld, E. (1989). *Constructivism in education.* Retrieved September 21, 2005, from http://www.univie.ac.at/constructivism/EvG/papers/114.pdf

Von Glasersfeld, E. (1997). *Piaget's legacy: Cognition as adaptive activity.* Retrieved October 11, 2005, from http://srri.nsm.umass.edu/vonGlasersfeld/onlinePapers/html/245.html

Voss, J. F., Greene, T. R., Post, T. A., & Penner, B. C. (1983). Problem-solving skill in the social sciences. In G. Bower (Ed.), *The psychology of learning and motivation* (Vol. 17, pp. 165–213). New York: Academic Press.

Voss, J. F., & Post, T. A. (1988). On the solving of ill-structured problems. In M. T. H. Chi, R. Glaser, & M. J. Farr (Eds.), *The nature of expertise* (pp. 261–285). Hillsdale, NJ: Lawrence Erlbaum Associates.

Walz, D. B., Elam, J. J., Krasner, H., & Curtis, B. (1987). A methodology for studying software design teams: An investigation of conflict behaviors in the requirements definition phase. In G. M. Olson, S. Sheppard, & E. Soloway (Eds.), *Empirical Studies of Programmers: Second Workshop (ESP2)* (pp. 83–99). Norwood, NJ: Ablex.

Warren, C., & Whitefield, A. (1987). The role of task characterization in transferring models of users: The example of engineering design. In H.-J. Bullinger & B. Shackel (Eds.), *Human–computer interaction-INTERACT'87.* Amsterdam: North-Holland.

Weber, G. (1991, August). *Explanation-based retrieval in a case-based learning model.* Paper presented at the Thirteenth Annual Meeting of the Cognitive Science Society, Evanston, IL.

Whitefield, A. (1986, September). *An analysis and comparison of knowledge use in designing with and without CAD.* Paper presented at CAD86, the Seventh International Conference on the Computer as a Design Tool, London.

Whitefield, A. (1989). Constructing appropriate models of computer users: The case of engineering designers. In J. Long & A. Whitefield (Eds.), *Cognitive ergonomics and human–computer interaction* (pp. 66–94). Cambridge, England: Cambridge University Press.

Wiedenbeck, S., & Scholtz, J. C. (Eds.). (1997). *Empirical Studies of Programmers: Seventh Workshop (ESP7).* New York: ACM Press.

Winograd, T. (Ed.). (1996). *Bringing design to software.* New York: ACM Press.

Winograd, T. (1997). *From computing machinery to interaction design.* Retrieved July 26, 2004, from http://hci.stanford.edu/winograd/acm97.html

Winograd, T., & Flores, F. (1986). *Understanding computers and cognition: A new foundation for design.* Norwood, NJ: Ablex.

Wirth, N. (1971). Program development by stepwise refinement. *Communications of the ACM, 14*(4), 221–227.

Yourden, E., & Constantine, L. (1978). *Structured design.* New York: Yourden Press.

Zhang, J. (1997). The nature of external representations in problem solving. *Cognitive Science*, *21*(2), 179–218.

Zhang, J., & Norman, D. A. (1994). Representations in distributed cognitive tasks. *Cognitive Science*, *18*, 87–122.

Author Index

255

Subject Index